GREEN CARBON
THE ROLE OF NATURAL FORESTS IN CARBON STORAGE

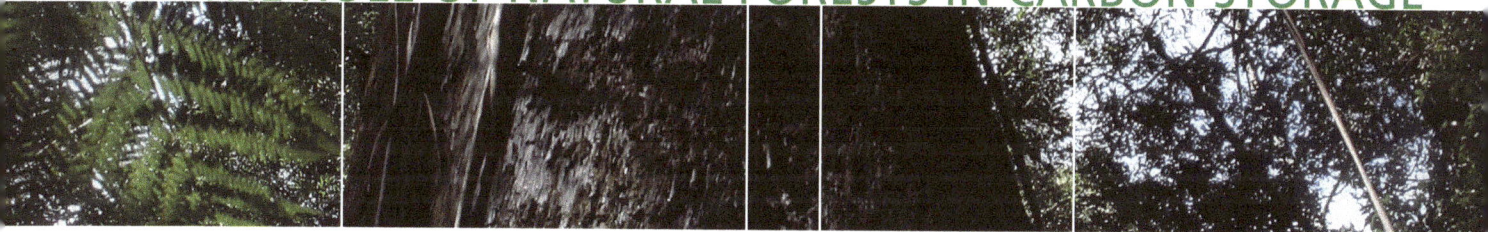

PART 1. A GREEN CARBON ACCOUNT OF AUSTRALIA'S SOUTH-EASTERN EUCALYPT FORESTS, AND POLICY IMPLICATIONS

Brendan G. Mackey, Heather Keith, Sandra L. Berry and David B. Lindenmayer
The Fenner School of Environment & Society, The Australian National University

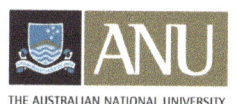

ANU
THE AUSTRALIAN NATIONAL UNIVERSITY

E PRESS

ANU
E PRESS

Published by ANU E Press
The Australian National University
Canberra ACT 0200, Australia
Email: anuepress@anu.edu.au
Web: http://epress.anu.edu.au
Online version available at: http://epress.anu.edu.au/green_carbon_citation.html

National Library of Australia Cataloguing-in-Publication entry

Title:	Green carbon : the role of natural forests in carbon storage. Part 1, A green carbon account of Australia's south-eastern Eucalypt forest, and policy implications / Brendan Mackey ... [et al.].
ISBN:	9781921313875 (pbk.) 9781921313882 (pdf.)
Subjects:	Carbon--Environmental aspects. Forests and forestry--Environmental aspects. Plants--Effect of atmospheric carbon dioxide on. Carbon dioxide mitigation.
Other Authors/Contributors:	
	Mackey, Brendan.
Dewey Number:	577.3144

This edition © 2008 ANU E Press

Design by ANU E Press

Cover photograph: *Eucalyptus regnans*, Central Highlands Victoria © 2007 Sarah Rees

TABLE OF CONTENTS

EXECUTIVE SUMMARY

Leaves: *E. delegatensis*, Bago State Forest, southern NSW. Photo: Claudia Keitel.

E. nitens, Erinundra, East Gippsland (620 t C ha⁻¹ of biomass carbon). Photo: Ern Mainka.

THE INTERNATIONAL COMMUNITY HAS NOW RECOGNIZED THE NEED FOR REDUCING EMISSIONS FROM DEFORESTATION AND FOREST DEGRADATION (REDD) AS A VITAL COMPONENT OF A COMPREHENSIVE SOLUTION TO THE CLIMATE CHANGE PROBLEM.

Only since the 2007 United Nations Climate Change Conference in Bali (UNFCCC CoP 13) have international negotiations focused on the role of natural forests in storing carbon.

The Intergovernmental Panel on Climate Change (IPCC) has identified the need for forest-based mitigation analyses that account for natural variability, that use primary data and that provide reliable baseline carbon accounts. In response, we are conducting a series of investigations into the carbon stocks of intact natural forests over large geographical areas, inclusive of environmental factors operating at landscape and regional scales. We are also considering the carbon impacts of land-use activities, including commercial logging. The key question we are asking in our research is 'How much carbon can natural forests store when undisturbed by intensive human land-use activity?'

This report presents a summary of results from case studies in the eucalypt forests of south-eastern Australia. We use these results to frame a discussion of REDD and we make policy recommendations to help promote a scientific understanding of the role of natural forests in the global carbon cycle and in solving the climate change problem.

IN UNDERSTANDING THE ROLE OF NATURAL FORESTS IN THE GLOBAL CARBON CYCLE, AND CLIMATE CHANGE MITIGATION POLICIES, THE COLOUR OF CARBON MATTERS.

It is the biological, ecological and evolutionary dimension that distinguishes the 'green' carbon in natural forests from the 'brown' carbon of industrialized forests, especially monoculture plantations. Drawing on the same poetic licence, we refer to the inorganic carbon in the atmosphere (carbon dioxide) and the oceans (carbonate) as 'blue' carbon.

Natural forests are more resilient to climate change and disturbances than plantations because of their genetic, taxonomic and functional biodiversity. This resilience includes regeneration after fire, resistance to and recovery from pests and diseases, and adaptation to changes in radiation, temperature and water availability (including those resulting from global climate change). While the genetic and taxonomic composition of forest ecosystems changes over time, natural forests will continue to take up and store carbon as long as there is adequate water and solar radiation for photosynthesis.

The green carbon in natural forests is stored in a more reliable stock than that in industrialized forests, especially over ecological time scales. Carbon stored in industrialized forests has a greater susceptibility to loss than that stored in natural forests. Industrialized forests, particularly plantations, have reduced genetic diversity and structural complexity, and therefore reduced resilience to pests, diseases and changing climatic conditions.

The carbon stock of forests subject to commercial logging, and of monoculture plantations in particular, will always be significantly less on average (~40 to 60 per cent depending on the intensity of land use and forest type) than the carbon stock of natural, undisturbed forests. The rate of carbon fixation by young regenerating stands is high, but this does not compensate for the smaller carbon pools in the younger-aged stands of industrialized forests compared with those of natural forests. Carbon accounts for industrialized forests must include the carbon emissions associated with land use and associated management, transportation and processing activities.

AUSTRALIAN NATURAL FORESTS HAVE FAR LARGER CARBON STOCKS THAN IS RECOGNIZED.

Our analyses showed that the stock of carbon for intact natural forests in south-eastern Australia was about 640 t C ha^{-1} of total carbon (biomass plus soil, with a standard deviation of 383), with 360 t C ha^{-1} of biomass carbon (living plus dead biomass, with a standard deviation of 277). The average net primary productivity (NPP) of these natural forests was 12 t C ha^{-1} yr^{-1} (with a standard deviation of 1.8). The highest biomass carbon stocks, with an average of more than 1200 t C ha^{-1} and maximum of over 2000 t C ha^{-1}, are in the mountain ash (*Eucalyptus regnans*) forest in the Central Highlands of Victoria and Tasmania. This is cool temperate evergreen forest with a tall eucalypt overstorey and dense *Acacia* spp. and temperate-rainforest tree understorey.

CARBON-ACCOUNTING MODELS MUST BE CAREFULLY CALIBRATED WITH APPROPRIATE ECOLOGICAL FIELD DATA IN ORDER TO GENERATE RELIABLE ESTIMATES FOR NATURAL FORESTS.

Access to appropriate ecological field data is critical for accurate carbon accounting in natural forests, as otherwise erroneous values will be generated. Models must be designed and calibrated to reflect the fact that the carbon dynamics of natural forests are significantly different to those of industrialized forests, especially monoculture plantations. Among other things, the carbon in natural forests has a longer residence time. We demonstrated this point by comparing our data with values of forest carbon accounts calculated from two commonly referenced sources.

In terms of global biomes, Australian forests are classified as temperate forests. The IPCC default values for temperate forests are a carbon stock of 217 t C ha^{-1} of total carbon, 96 t C ha^{-1} of biomass carbon, and a NPP of 7 t C ha^{-1} yr^{-1}. The IPCC default values for total carbon are approximately one-third, and for biomass carbon approximately one-quarter that of the average values for south-eastern Australian eucalypt forests, and one-twentieth of the most biomass carbon dense eucalypt forests. We calculate the total stock of carbon that can be stored in the 14.5 million ha of eucalypt forest in our study region is 9.3 Gt[1], if it is undisturbed by intensive human land-use activities; applying the IPCC default values would give only 3.1 Gt.

The difference in carbon stocks between our estimates and the IPCC default values is the result of us using local data collected from natural forests not disturbed by logging. Our estimates therefore reflect the *carbon carrying capacity* of the natural forests. In heavily disturbed forests, the current carbon stocks reflect land-use history. The difference between the two is called the *'carbon sequestration potential'*—the maximum carbon stock that can be sequestered as the forest re-grows.

We also tested the Australian Government's National Carbon Accounting System (NCAS) (Australian Greenhouse Office 2007a) and found it underestimated the carbon carrying capacity of natural forests with high biomass stocks. NCAS was designed to model biomass growth in plantations and afforestation/reforestation projects using native plantings. The empirically based functions within NCAS were calibrated using data appropriate for that purpose. But, this meant that NCAS was unable to accurately estimate the carbon carrying capacity of carbon dense natural forests in south eastern Australia. However, the kinds of field data used in our study could be used to recalibrate NCAS so that it can generate reliable estimates of biomass carbon in these forests.

THE REMAINING INTACT NATURAL FORESTS CONSTITUTE A SIGNIFICANT STANDING STOCK OF CARBON THAT SHOULD BE PROTECTED FROM CARBON EMITTING LAND-USE ACTIVITIES.

THERE IS SUBSTANTIAL POTENTIAL FOR CARBON SEQUESTRATION IN FOREST AREAS THAT HAVE BEEN LOGGED IF THEY ARE ALLOWED TO RE-GROW UNDISTURBED BY FURTHER INTENSIVE HUMAN LAND-USE ACTIVITIES.

Our analysis shows that in the 14.5 million ha of eucalypt forests in south-eastern Australia, the effect of retaining the current carbon stock (equivalent to 25.5 Gt CO_2 (carbon dioxide)) is equivalent to avoided emissions of 460 Mt[2] CO_2 yr^{-1} for the next 100 years. Allowing logged forests to realize their sequestration potential to store 7.5 Gt CO_2 is equivalent to avoiding emissions of 136 Mt CO_2 yr^{-1} for the next 100 years. This is equal to 24 per cent of the 2005 Australian net greenhouse gas emissions across all sectors; which were 559 Mt CO_2 in that year.

---

1   Gigatonne (Gt) equals one billion or 1.0×10^9 tonnes.

2   Megatonne (Mt) equals one million or 1.0×10^6 tonnes

If, however, all the carbon currently stored in the 14.5 million ha of eucalypt forest in south-eastern Australia was released into the atmosphere it would raise the global concentration of carbon dioxide by 3.3 parts per million by volume (ppmv). This is a globally significant amount of carbon dioxide; since 1750 AD, the concentration of carbon dioxide in the atmosphere has increased by some 97 ppmv.

REDUCING EMISSIONS FROM DEFORESTATION AND FOREST DEGRADATION (REDD) IS IMPORTANT IN ALL FOREST BIOMES — BOREAL, TROPICAL AND TEMPERATE — AND IN ECONOMICALLY DEVELOPED AS WELL AS DEVELOPING COUNTRIES.

From a scientific perspective, green carbon accounting and protection of the natural forests in all nations should become part of a comprehensive approach to solving the climate change problem. Current international negotiations are focussed on reducing emissions from deforestation and forest degradation in developing countries only. However, REDD is also important in the natural forests of countries such as Australia, Canada, the Russian Federation, and the USA.

Part of the ongoing international climate change negotiations involves debate on the technical definition of key terms. 'Forest degradation' should be defined to include the impacts of any human land-use activity that reduces the carbon stocks of a forested landscape relative to its natural carbon carrying capacity. The definition of 'forest' should also be revised to recognize the differences between the ecological characteristics of natural forests and industrialized forests, especially plantations. These differences include the higher biodiversity, ecosystem resilience, and carbon residence time of natural forests.

E. regnans, Dandenong Ranges National Park, Victoria (900 t C ha^{-1} of biomass carbon). Photo: Sandra Berry.

INTRODUCTION

Natural forests play a significant role in the global carbon cycle. Biomass and soil store approximately three times the amount of carbon that is currently found in the atmosphere, and the annual exchange of carbon between the atmosphere and natural forests is 10 times more than the annual global carbon emissions from humans burning fossil fuels. Despite natural forests storing such significant amounts of carbon, to date there has been scant consideration given by policymakers to the role of forests in addressing the climate change problem. At the 2007 United Nations Climate Change Conference in Bali (UNFCCC CoP 13), however, the international community recognized the need to reduce emissions from deforestation and forest degradation (REDD) as a vital component of a comprehensive solution to the climate change problem.

The significance to the climate change problem of achieving REDD can be appreciated when we consider that about 35 per cent of greenhouse gases stored in the atmosphere is due to past deforestation, and about 18 per cent of annual global emissions is the result of continuing deforestation (IPCC 2007). Furthermore, even when forest is not cleared to make way for other land uses, there are significant and continuing emissions of carbon dioxide from commercial logging and other land-use activities that reduce the stock of carbon stored in the ecosystem. Consequently, there is now great interest in, and indeed an urgent need to develop and apply, methods that better quantify the carbon stored in natural forests and how these pools change as the result of human land-use activities.

While international attention is now focused on REDD in developing countries, the laws of nature that account for the global carbon cycle operate irrespective of political boundaries. Therefore, a unit of carbon emitted due to deforestation and forest degradation in Australia, the United States, Canada or Russia has exactly the same impact on atmospheric greenhouse gas levels as a unit of carbon emitted from deforestation and degradation of forests in Indonesia, Papua New Guinea, the Congo Basin or Brazil. From a scientific perspective, solving the climate change problem requires, among others things, that REDD be accounted for in all forest biomes, irrespective of the host nation's economic status.

The Intergovernmental Panel on Climate Change (IPCC) has identified the need for forest-based mitigation analyses that account for natural variability, use primary data and provide reliable baseline carbon accounts (Nabuurs et al. 2007). In response, we are conducting a series of investigations into the carbon stocks of intact natural forests over large geographical areas, inclusive of environmental factors operating at landscape and regional scales. We are also considering the carbon impacts of land-use activities, including commercial logging.

In Australia, a number of studies have examined carbon stocks at continental scales (Barrett 2002) and using fine-resolution land-cover data (Brack et al. 2006). There is, however, a lack of baseline carbon accounts for natural forests undisturbed by intensive human land-use activities. Such baselines are essential if we are to value accurately the carbon stored in natural forests, and in order to account properly for the carbon emissions from land-use activities.

An approach to estimating the carbon stocks of intact natural forests was developed and tested by Roxburgh et al. (2006). Our study extends this approach by applying it over entire regions. The approach is based on estimating what we call the 'natural carbon carrying capacity' of a landscape. The natural carbon carrying capacity is defined as the mass of carbon able to be stored in a forest ecosystem under prevailing environmental conditions and natural disturbance regimes, but excluding disturbance by human activities (Gupta and Rao 1994). This estimate provides an appropriate baseline for estimating the impacts on carbon stocks of intensive human land-use activities. Once the natural carbon carrying capacity is established, it is possible to calculate the potential increase in carbon storage that would occur if land-use management were changed and carbon-emitting land-use ceased. This potential increase in the carbon stored in the forest is called the 'carbon sequestration potential'.

The key question we are asking in our research is 'How much carbon can natural forests store when undisturbed by intensive human land-use activity?' This report presents a summary of results from case studies in the eucalypt forests of south-eastern Australia. We use these results to frame a discussion of REDD and we make policy recommendations to help promote a scientific understanding of the role of natural forests in the global carbon cycle and in solving the climate change problem.

This report was prepared in response to the considerable public interest in the issue of REDD. An earlier version was written as preparatory material for the Bali 2007 Climate Change Conference. A technical paper that details the source data, the methods used and the full results is being prepared for a scientific journal. In the interim, any technical questions regarding data and methods should be directed to the authors.

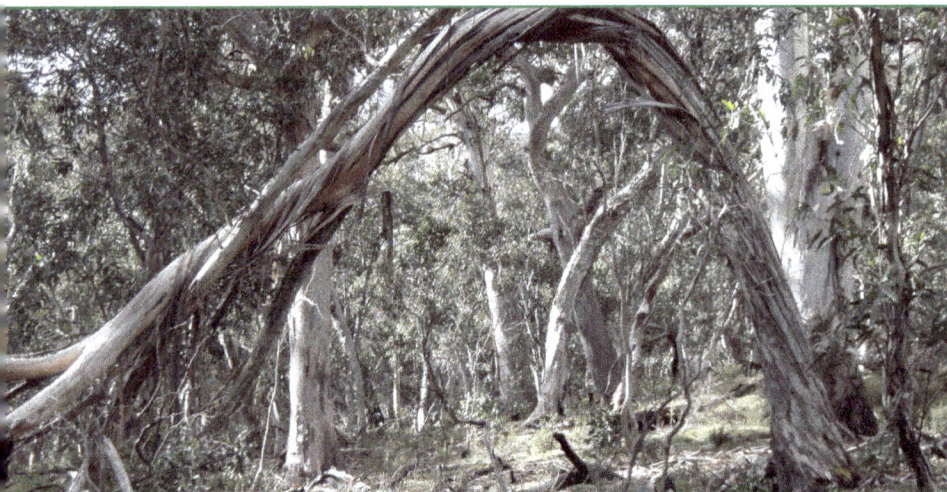

E. dalrympleana, E. pauciflora, subalpine forest, Kosciuszko National Park, NSW (325 tC ha^{-1} of biomass carbon). Photo: Ian Smith.

THE SIGNIFICANCE OF GREEN CARBON

Decorticating bark: *E. dalrympleana*, Kosciuszko National Park. Photo: Ian Smith.

WHAT IS GREEN CARBON?

It is useful to consider the 'colour' of carbon when considering the role of natural forests in the global carbon cycle.

Grey carbon[3] is the carbon stored in fossil fuel (coal, oil and gas deposits in the lithosphere).

Green carbon is the carbon stored in the biosphere. We call it 'green' because carbon is taken up from the atmosphere by plants through the process of photosynthesis, which is dependent on the green chlorophyll pigment found in plant leaves[4]. Here, we use the term green carbon to refer to the carbon sequestered through photosynthesis and stored in natural forests. Natural forests are defined here as forests that have not been disturbed by intensive human land-use activities, including commercial logging.

Brown carbon is the carbon stored in industrialized forests. These are forests that are logged commercially for their wood, which is used as a source of raw material for industrial manufacturing processes. There are two types of industrialized forests: 1) where tree regrowth is from the naturally occurring tree stock and seed bank; and 2) where the trees are planted by humans and usually comprise a single tree species, much like a monoculture crop. Industrialized forests constitute a stock of organic carbon and are therefore part of the biosphere; however, we consider this carbon to be 'brown' in colour rather than 'green' in order to stress the fact that industrialized forests are a 'mix' of green and grey carbon[5]. Fossil fuel is expended and therefore grey carbon emitted in managing these forestry operations and from the associated industrial processes.

Blue carbon refers to the inorganic carbon stored in the atmosphere (carbon dioxide, CO_2) and oceans (carbonate, CO_3^{2-}). While there are significant stocks of marine green carbon in the ocean[6], here we are concerned with the green carbon stored in terrestrial ecosystems, and natural forests in particular.

The significance of natural forests to mitigating the climate change problem is a hotly debated topic. Some commentators argue that forest protection is a secondary issue and the primary focus of discussion should be on approaches to reducing emissions of grey carbon from burning fossil fuels. We can, however, no longer afford the luxury of ignoring any one of the components of the

---

3   In greenhouse literature, the term 'black carbon' has been used to refer to charcoal in soil and soot in the atmosphere.

4   Carbon is taken up from the atmosphere by photosynthesising bacteria and algae, in addition to plants.

5   We have of course taken some poetic licence in using these colours to describe the different states of carbon. The colour brown is in reality produced from a mix of the three primary colours and not from simply mixing green and grey.

6   There is also biological uptake in the oceans, but the carbon dioxide first physically dissolves from the atmosphere into the ocean, then the dissolved inorganic carbon can be taken up by photosynthesising phytoplankton.

global carbon cycle that are being disrupted by human activity.

Solving the climate change problem requires that atmospheric concentrations of greenhouse gases be reduced and stabilized to a level that prevents dangerous anthropogenic interference with the climate system (UNFCCC). What constitutes a 'safe level' is a critical question that is being debated actively among scientists and policy advisors. Evidence from glacial ice cores has revealed that atmospheric concentrations of carbon dioxide ranged between 180 and 300 parts per million by volume (ppmv) in the past 650 000 years (with typical maximum values of 290 ppmv) (Petit et al. 1999; IPCC 2007). Assuming this natural variability revealed by the ice-core records persisted[7], we should assume a maximum safe level is 300 ppmv. In the language of thermodynamics: through the interactions of various natural processes, Earth's average planetary temperature has been maintained in a state of dynamic equilibrium in the past 650 000 years where the temperature varies but within a well-defined 'ceiling' and 'floor'.

As a result of humans burning fossil fuels and causing emissions from deforestation and forest degradation (especially in the past 100 years), the current level of atmospheric carbon dioxide is about 380 ppmv (IPCC 2007). We have therefore already exceeded a safe level of atmospheric carbon dioxide as defined by the natural variability of the past 650 000 years. Stabilizing atmospheric carbon dioxide at between 350 and 400 ppmv will require that emissions are reduced to approximately 85 per cent of 2000 levels by 2050, and that the peak year for emissions is not later than 2015 (IPCC 2007). Meeting this target will still result in a projected temperature increase of 2 to 2.4°C and a sea-level rise of 0.4 to 1.4 m. Given the current trajectory of emissions, the scientific community is now discussing the consequences of atmospheric levels of carbon dioxide reaching up to 790 ppmv by 2100 (IPCC 2007), which is predicted to result in a temperature increase of up to 6.1°C and a sea-level rise of 1.0 to 3.7 m.

We can no longer afford to ignore emissions caused by deforestation and forest degradation from every biome (that is, we need to consider boreal, tropical and temperate forests) and in every nation (whether economically developing or developed). We need to take a fresh look at forests through a carbon and climate change lens, and reconsider how they are valued and what we are doing to them.

7   The ice-core records confirm that the Earth has experienced a long sequence of cool and warm periods associated with oscillations in the planetary orbit around the sun. A very long cooling phase (about 100 000 years) culminates in a glacial maximum followed by a rapid warming to reach a temperature maximum (about 10 000 years) (Berger and Loutre 2002).

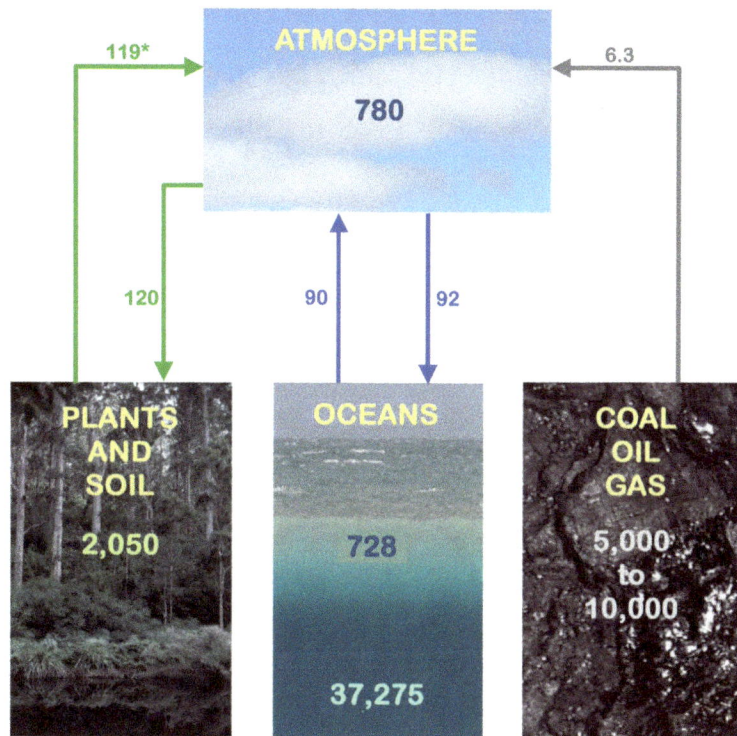

FIGURE 1: GLOBAL CARBON CYCLE

Approximate global carbon cycle stocks (boxes) and fluxes (arrows). (Adapted from Houghton 2007). Units are Gt of carbon, and fluxes are per year. The colours of the arrows correspond to the definition of colour of carbon. * Deforestation contributes ~2 Gt C yr⁻¹.

WHAT IS THE ROLE OF FORESTS IN THE CARBON CYCLE?

Terrestrial ecosystems—especially natural forests—play a critical role in regulating greenhouse gas concentrations in the atmosphere and therefore must be part of a comprehensive response to the climate change problem. An appreciation of the significance of natural forests in the carbon cycle requires understanding of how Earth functions as a system. Because Earth is a closed system in terms of chemical elements, the atomic components of the major greenhouse gases (water vapour, carbon dioxide and methane) are neither created nor destroyed. Rather, they reside in and move between reservoirs (also called 'stocks' or 'pools') within the global carbon and hydrological cycles. As they move between reservoirs, carbon and water change state: water from a liquid, to gas or ice; and carbon from inorganic gases in the atmosphere, to organic compounds in living and dead organisms on land and in the sea, to inorganic substances in the oceans and the Earth's crust.

A simplified diagram of the global carbon cycle is shown in Figure 1. The estimates of global carbon stocks and fluxes are only approximate due to lack of data. The annual uptake of carbon (as carbon dioxide) by plants (through photosynthesis) from the atmosphere to the plant and soil reservoir (organic carbon) is about 120 Gt yr⁻¹. Through the respiration of living organisms (including humans and their livestock), and oxidative combustion by fire, a little less than 120 Gt yr^{-1} of 'plant and soil' reservoir carbon is emitted to the atmosphere. The biosphere is estimated to be a small carbon sink. Approximately 2 Gt of carbon emissions by the plant and soil reservoir is due to deforestation. This acts to increase the loss of carbon and decrease the uptake of carbon by the plant and soil reservoir. Over time, therefore, the size of the reservoir of carbon in plants and the soil is decreasing. The coal/oil/gas reservoir (which supplies most of the energy requirements of industry) is also decreasing by approximately 6 Gt yr^{-1}. If there is less carbon in the plant and soil pool, and in the coal/oil/gas pool, there must be more in the atmospheric and ocean pools. To date, humans have released about 300 Gt of grey carbon, but there is over 5000 Gt remaining in the lithosphere that potentially can be accessed for human use (Archer 2005). About 2000 Gt of carbon is estimated to reside currently in terrestrial ecosystems (plant and soil reservoirs), with about 75 per cent of this stored in natural forest ecosystems. However, about 50 per cent of the world's forests have been cleared so that current terrestrial carbon stocks are substantially below their natural carbon carrying capacity (Archer 2005; MEA 2005; Houghton 2007).

Carbon cycles between the lithosphere, hydrosphere, atmosphere and biosphere, but its residence time in each of these reservoirs varies significantly[8]. The concentration of carbon in the atmosphere due to the release of carbon from the lithosphere reservoir will remain at elevated levels for a long time even if grey carbon emissions are stopped immediately (Archer 2005). The two pathways for transfer of carbon out of the atmosphere are: 1) dissolution in river and ocean water and, eventually, incorporation into carbonate rock; and 2) uptake of carbon by plants and storage in the biosphere. The terrestrial biosphere–atmosphere fluxes operate on a faster time scale and are under a greater degree of human control than the fluxes of the hydrosphere. Solving the climate change problem will require both reducing grey-carbon emissions and maximising the uptake of carbon in the biosphere. A healthy biosphere provides a buffering capacity for changes in the carbon cycle.

ARE GREEN CARBON STOCKS RELIABLE?

The argument is commonly heard that forests are an unreliable carbon sink because of their vulnerability to fire, pests, diseases and drought, which can reduce the standing stock of carbon and inhibit forest growth. Another argument is that climate change might cause conditions to be less conducive to forest growth, for example, by reducing water available for photosynthesis or increasing temperatures beyond the thermal tolerance of tree species, thereby causing forests to become a source of rather than a sink for carbon. It is also argued that the stock of green carbon

---

8   Residence time is the average time a unit of carbon spends in a given reservoir, that is, carbon stock or pool. It is calculated by dividing the reservoir volume by the rate of flow.

is too small to make a significant contribution to greenhouse gas mitigation and is therefore not an important policy consideration.

As noted earlier, green carbon in the biosphere has a significantly different residence time compared with grey carbon in the lithosphere. Therefore, in terms of the global carbon cycle, green and grey carbon should not be treated as equivalent with respect to policy options. In terms of preventing harmful change to the climate system, it is important to avoid emissions of grey carbon from burning fossil fuels, and leave oil, gas and coal stored in the lithosphere. Additionally, the uptake and storage of carbon by natural forests has a powerful and relatively rapid negative feedback on the enhanced greenhouse effects from emissions. Feedbacks are the key to understanding how relatively minor increases in greenhouse gas concentrations can result in massive changes in Earth's climate system (Hansen et al. 2007).

Generally, a greenhouse-enhanced world is a warmer and wetter world—albeit with changing regional patterns (Zhang et al. 2007). Water is essential for photosynthesis (the uptake of carbon by plants from the atmosphere) and production of new biomass. When water is plentiful (and the soil is not degraded), atmospheric carbon will continue to be sequestered in new biomass. In addition, as atmospheric levels of carbon dioxide increase, photosynthesis becomes more efficient as plants can fix more carbon dioxide using the same amount of water (Farquhar 1997). Increased cloud cover (associated with increased rainfall) is not necessarily an impediment as photosynthesis utilizes diffuse as well as direct solar energy (Farquhar and Roderick 2003), and it could even enhance photosynthesis in multi-layered vegetation canopies (Hollinger et al. 1998).

The stock of green carbon in an ecosystem is the result of the difference between the rates of biomass production and decomposition. Like the global carbon cycle, green carbon cycles between pools: living biomass, dead biomass and soil. The residence time of a unit of carbon in each pool varies—the longest is for woody biomass and soil (Roxburgh et al. 2006). Rates of decomposition scale with increasing temperature and moisture (Golley 1983). An excess of soil water, however, leads to anaerobic conditions, a decrease in decomposition and a build-up of dead organic matter. This is why tropical peat forests and boreal forests have large pools of soil organic carbon, while tropical and temperate forests have proportionally more living biomass carbon.

Various processes enable forests to persist in the face of changing environmental conditions, including climate change. Natural forests are characterized by a rich biodiversity at all levels: genetic, taxonomic and ecosystem. This is obvious especially when, in addition to the diversity of plants and vertebrate animals, we consider the invertebrates, bacteria and fungi, and the vast webs of ecological and coevolving interactions that together constitute a functioning ecosystem (Odum and Barret 2005; Thompson 2005). The genetic diversity found within species provides the capacity for, among other things, micro-evolution whereby populations can become rapidly adapted to local conditions (Bradshaw and Holzapfel 2006). High taxonomic diversity provides a pool of species with different life histories and niche tolerances from which natural selection can reveal the plant or animal best suited to new conditions (Hooper et al. 2005). Natural selection, acting on the rich biodiversity found in natural forests, can also result in the optimisation of plants' physiological processes (Cowan and Farquhar 1977) and in the optimization of trophic interactions (Brown et al. 2004) in response to environmental change. Natural forests are therefore more resilient to climate change and disturbances than plantations because of their genetic, taxonomic and functional biodiversity. This resilience includes regeneration after fire, resistance to and recovery from pests and diseases and adaptation to changes in radiation, temperature and water availability.

Oxygenic/photosynthetically based ecosystems have persisted on Earth for at least 2.8 billion years (Des Marais 2000), due in no small measure to the kinds of biological, ecological and evolutionary processes noted above. While the genetic and taxonomic composition of forest ecosystems changes over time, forests will continue to uptake and store carbon as long as there is adequate water and solar radiation for photosynthesis. From this perspective, the carbon in natural forests is stored in a more reliable stock than in industrialized forests, especially over ecological time scales. Carbon stored in industrialized forests has a greater susceptibility to loss than that stored in natural forests. Regrowth forests and plantations have reduced genetic diversity and structural complexity, and therefore reduced resilience to pests, diseases and changing climate conditions (Hooper and Vitousek 1997; Hooper et al. 2005, McCann 2007). The risk of fire in industrialized forests is greater than in natural forests because of the associated increase in human activity in the area, the use of machinery and public access.

Given the resilience of natural ecosystems, the green carbon stocks in forest biomes are more likely in the longer term to expand than to shrink under enhanced greenhouse conditions, and in the absence of perturbations from human land-use activities[9]. Indeed, the negative feedback (with respect to increased atmospheric concentrations of carbon dioxide) provided by enhanced plant

E. obliqua, Mt. Wellington, Tasmania.
Photo: Rob Blakers.

9   This statement must, however, be qualified by the high level of uncertainty about regionally scaled climate change predictions of rainfall and evaporation—the main variables controlling water availability.

growth has been argued to be critical to the long-term stability of Earth's environment within the bounds conducive to life (Gorshkov et al. 2000).

WHAT ABOUT INDUSTRIALIZED FORESTS?

There are important distinctions between the carbon dynamics of natural forests and industrialized forests, especially monoculture plantations. The majority of biomass carbon in natural forests resides in the woody biomass of large old trees. Commercial logging changes the age structure of forests so that the average age of trees is much younger. The result is a significant (more than 40 per cent) reduction in the long-term average standing stock of biomass carbon compared with an unlogged forest (Roxburgh et al. 2006; Brown et al. 1997). Plantations are designed to have all of their above-ground biomass removed on a regular basis. The rotation period between harvests varies from 10 to 70 years globally, depending on species and commercial purposes (Varmola and Del Lungo 2003). The carbon stock of forests subject to commercial logging—and of monoculture plantations in particular—will therefore always be significantly less on average than the carbon stock of natural, undisturbed forests.

It is argued by some industry advocates that commercial logging is greenhouse gas neutral because: a) young trees have high rates of growth and carbon fixation; and b) some of the biomass removed from the forest is used for wood-based products with a substantial residence time. Regarding the first point, it is true that the rate of carbon uptake by young trees in plantations and regrowth forests is high. However, this carbon uptake over a rotation would not compensate for the amount of carbon presently stored in natural forests that would be lost if they were harvested (Harmon et al. 1990; Schulze et al. 2000). Responding to the second point, it is critical from a carbon-mitigation perspective to account for all carbon gains and losses associated with logging and associated industrial processes. Comprehensive carbon accounting is needed that includes carbon uptake and emissions from all human activities associated with commercial logging and processing of the associated wood-based products, as well as carbon storage in products.

Emissions that need to be accounted for include grey carbon from burning fossil fuels for energy to do work and green carbon from killing living biomass and accelerating the rate of decomposition of dead biomass. When considering the carbon accounts associated with industrialized forests, it is therefore necessary to include carbon emissions resulting from: a) forest management (for example, the construction and maintenance of roads, post-logging regeneration burns); b) harvesting (including use of machinery, and wastage from collateral damage to living woody biomass and soil carbon); c) transportation of logs, pulpwood and woodchips; and d) manufacturing. All of these emissions must be subtracted from the carbon stored in wood-based products. Also, it needs to be demonstrated that the carbon in wood-based products will remain in the terrestrial biosphere carbon reservoir for a longer period than it would have if it had remained in an unlogged natural forest.

Ideally, a comprehensive carbon audit should be conducted using the energy audit method of Odum (1981). We cannot find any such comprehensive accounts of the grey carbon emitted from commercial logging and wood-products manufacturing inclusive of all stages in the product life cycle: forest management, harvesting, transportation and manufacturing. Of these, the most critical are likely to be: 1) collateral damage to forest biomass and soil carbon (also called 'wastage'); and 2) the differences between the residence time of carbon in the natural forest pools and the wood-product pools. In natural forests with large carbon stocks, the wastage of biomass due to commercial logging is significant. For example, commercial logging in tropical natural forests has been shown to dramatically reduce carbon stocks. In Papua New Guinea, commercial logging has been found to result in about 27 per cent of stem volume being removed, another 13 per cent being killed and half of the trees with a stem diameter of more than 5 cm destroyed (Abe et al. 1999). The residence time of the wood-based products is also a critical factor given the longevity of woody stems, coarse woody debris and soil carbon pools in natural forest (Roxburgh et al. 2006). An additional critical consideration is the loss of green carbon from natural forest pools when industrialized forests and plantations are first established, and the time it will take for this biomass to be regrown (Fargione et al. 2008).

In summary, forest protection is an essential component of a comprehensive approach to mitigating the climate change problem for a number of key reasons. These include:

- For every hectare of natural forest that is logged or degraded, there is a net loss of carbon from the terrestrial carbon reservoir and a net increase of carbon in the atmospheric carbon reservoir. The resulting increase in atmospheric carbon dioxide exacerbates climate change.

- Given the long time that grey carbon will remain in the atmosphere–biosphere–hydrosphere system, maintaining the natural processes that regulate atmosphere–biosphere fluxes will be critical for moderating carbon levels in the atmosphere in the short to medium term. If natural forests are able to expand then the increased buffering capacity will act as a negative feedback on the accumulation of greenhouse gasses.

- The carbon dynamics of natural forests are significantly different to those of industrialized forests, especially monoculture plantations. The carbon in natural forests has a longer residence time, the system is more resilient to environmental perturbations and natural processes enable ecological systems and their component species to respond to changing conditions.

THE GREEN CARBON BASELINE PROBLEM

In recognizing the importance of reducing emissions from deforestation and forest degradation (REDD), the international community is now exploring appropriate mechanisms that will provide the financial investments needed to protect natural forests and keep them intact. Irrespective of the mechanism, it will be essential to have reliable estimates of baseline carbon accounts against which changes in carbon stocks can be gauged. Two kinds of baselines are needed: 1) the current stock of carbon stored in forests; and 2) the natural carbon carrying capacity of a forest (the amount of carbon that can be stored in a forest in the absence of human land-use activity). The difference between the two is called the carbon sequestration potential—the maximum amount of carbon that can be stored if a forest is allowed to grow given prevailing climatic conditions and natural disturbance regimes.

The greater the carbon sequestration potential of a forest, the more the carbon stock has been degraded by human land-use activities. It follows that stopping the carbon-degrading land-use activities will allow the forest to regrow carbon stocks to their potential—assuming the natural regenerative capacity of the ecosystem is maintained. Most carbon accounting schemes focus simply on the current carbon stocks in a landscape and do not consider a forest's natural carbon carrying capacity. This is partly because the concept is not widely appreciated but also because its calculation is difficult.

It is not possible to predict the carbon carrying capacity of a natural forest reliably from process-based simulation models. This is because the carbon stock is the result of a complex set of multi-scaled natural processes, some of which can be modelled reliably (for example, gross primary productivity), while others cannot because they are understood only poorly (in particular, allocation of biomass components, turnover times of components and rates of decomposition). Consequently, estimating carbon carrying capacity relies on empirical data gathered from natural forests largely undisturbed by human land-use activity. Natural disturbances, however, have to be taken into account. As noted above, commercial logging significantly reduces the standing stock of carbon below the natural carbon carrying capacity because most of the biomass carbon in a forest is in the woody stems of large trees (more than 70 cm diameter at breast height; Brown et al. 1997), which are removed over time. In contrast, tree mortality by natural processes such as wind, fire or pests removes more of the small, weaker trees and a smaller proportion of large trees. The role of fire in natural forests is complex and must be considered on a landscape-wide basis in terms of the pattern of fire events over time (so-called 'fire regimes') (Mackey et al. 2002). It follows that estimating natural carbon carrying capacity requires data that sample the range of ecosystem conditions found in a natural forest.

Conventional approaches to estimating biomass carbon stocks are based on stand-level commercial forestry inventory techniques. These data are not, however, suitable for calculating the carbon carrying capacity of natural forests. In industrialized forests, mensuration is focused mainly on estimating regrowth rates in logged stands. Consequently, the most commonly available field-survey data about the standing crop of carbon in forests are from regrowth stands. These data cannot be used to estimate the carbon stocks of ecologically mature natural forests. To estimate the carbon carrying capacity of a natural forest, field data are needed from sites that have not been subjected to commercial logging and that sample all carbon pools in the ecosystem (living biomass, dead biomass and soil) at appropriate space/time scales. As natural forests can take 200 to 400 or more years to reach their mature biomass levels (Saldarriaga et al. 1988; Dean et al. 2003), carbon accounts must reflect such long-term dynamics.

In the next section, we present some results from our continuing investigations into baseline green carbon accounts using the eucalypt forests of south-eastern Australia as our case study. We present estimates of the natural carbon carrying capacity of these forest ecosystems. We then use these results to consider some of the policy implications for reducing emissions from deforestation and forest degradation.

Canopy leaves: Lamington National Park, Queensland. Photo: Heather Keith.

SOUTH-EASTERN AUSTRALIA EUCALYPT FOREST CASE STUDY

INTRODUCTION

The location of the study region is shown in Figure 2. Our approach draws on existing methods plus some innovations necessary to deal with various problems that arise, including: a) stand ages are often unknown and stands are commonly multi-aged; b) disturbance and land-use history might be unknown; c) forests that have remained undisturbed by human land-use activity usually occur in rugged topography; and d) little information exists about the growth curves over time of many tree species. Analyses drew on a range of inputs: remote sensing data, spatially explicit environmental variables and site data that sampled carbon pools.

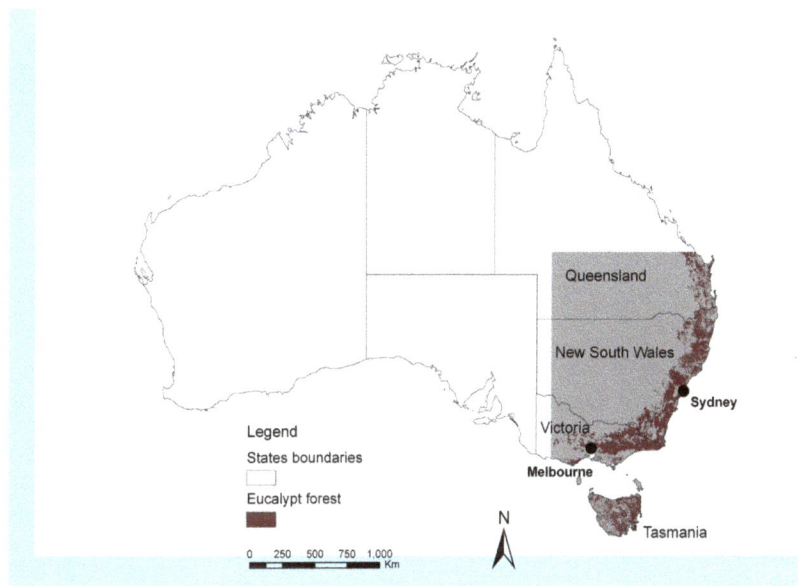

FIGURE 2: LOCATION OF THE CASE STUDY REGION, AND THE NATURAL EUCALYPT FORESTS IN SOUTH-EASTERN AUSTRALIA

The analytical framework developed to estimate the carbon carrying capacity of eucalypt forests in south-eastern Australia was based on knowledge of ecological processes as represented in Figure 3. Gross primary productivity (GPP) is the annual rate of carbon uptake by photosynthesis. Net primary productivity (NPP) is the annual rate of carbon accumulation in plant tissues after deducting the loss of carbon dioxide by autotrophic (plant) respiration (R_a). This carbon is used for production of new biomass components—leaves, branches, stems, fine roots and coarse roots—which increments the carbon stock in living plants. Mortality and the turnover time of carbon in these components vary from weeks (for fine roots), months or years (for leaves, bark and twigs) to centuries (for woody stem tissues). Mortality produces the dead biomass components that provide the input of carbon to the litter layer and soil through decomposition. The carbon that is consumed by herbivores and micro-organisms is emitted as carbon dioxide to the atmosphere by the process of heterotrophic respiration (R_h). The remaining carbon contributes to accumulation in the soil. Accumulation

CARBON POOLS IN NATURAL FORESTS

(pools as an average per cent of total carbon stock)

Living aboveground biomass (43%): *Corymbia maculata*, south coast NSW. Photo: Sandra Berry.

Litter layer (2%): *E. fastigata* forest, Shoalhaven catchment. Photo: Sandra Berry.

Dead biomass in stags (6%): *E. regnans*, central highlands, Victoria. Photo: Luke Chamberlain.

Root biomass (8%): *E. delegatensis*, Bago State Forest, southern NSW. Photo: Heather Keith.

Coarse woody debris (7%): *E. obliqua*, Mt. Wellington, Tasmania. Photo: Rob Blakers.

Soil profile (34%): Red Dermosol, Bago State Forest, southern NSW. Photo: Heather Keith.

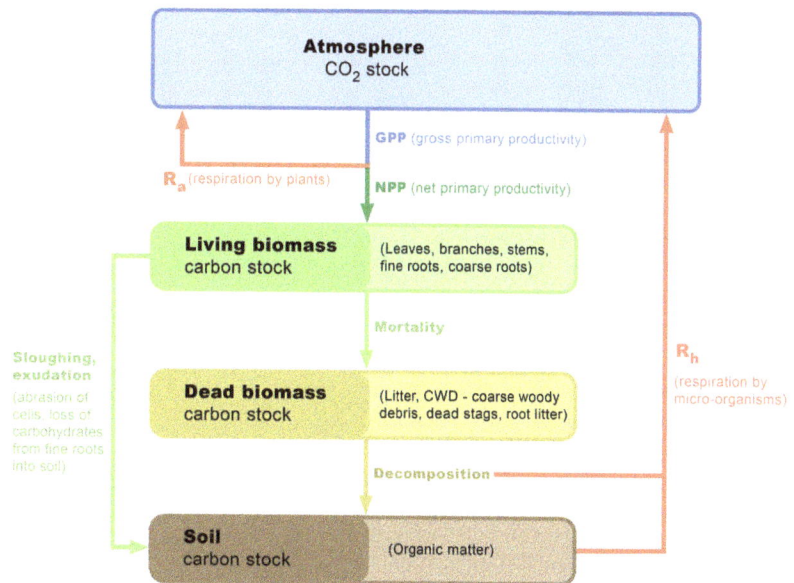

FIGURE 3: FRAMEWORK ILLUSTRATING THE ECOLOGICAL PROCESSES INVOLVED IN ESTIMATING THE CARBON CARRYING CAPACITY OF NATURAL FORESTS (THAT IS, GREEN CARBON STOCKS).

Boxes represent stocks of carbon, and arrows represent fluxes (movement) of carbon.

of carbon in the plant and soil reservoir is highly dependent on the residence time of each of the components of living and dead biomass and soil. Little information about these processes exists for natural forests. Therefore, our empirical approach to estimate carbon carrying capacity used site-specific data from natural forests largely undisturbed by human land-use activity.

The outcome of our analyses was an estimate of the carbon carrying capacity of the natural eucalypt forests in south-eastern Australia[10], which are shown in Figure 2. Analyses were restricted to forested land with environmental conditions that were within the numerical ranges sampled by our site data—yielding an area of approximately 14.5 million ha.

SUMMARY OF METHODS

Gross primary productivity (GPP) was calculated using the method of Roderick et al. (2001), as applied by Berry et al. (2007; see also Mackey et al. 2008). The source data were a continental time series of GPP modelled from the NASA MODIS (MOD13Q1) satellite data (Barrett et al. 2005) at a resolution of 250 m. The value of GPP used was the maximum annual value for the period from 1 July 2000 to 30 June 2005 (the maximum was used in order to exclude periods of major disturbance such as the 2003 bushfires).

---

10   These forests were defined as Major Vegetation Groups 2 and 3 in the National Vegetation Information System (NVIS 2003), where tree height is greater than 10 m and canopy cover is greater than 30%.

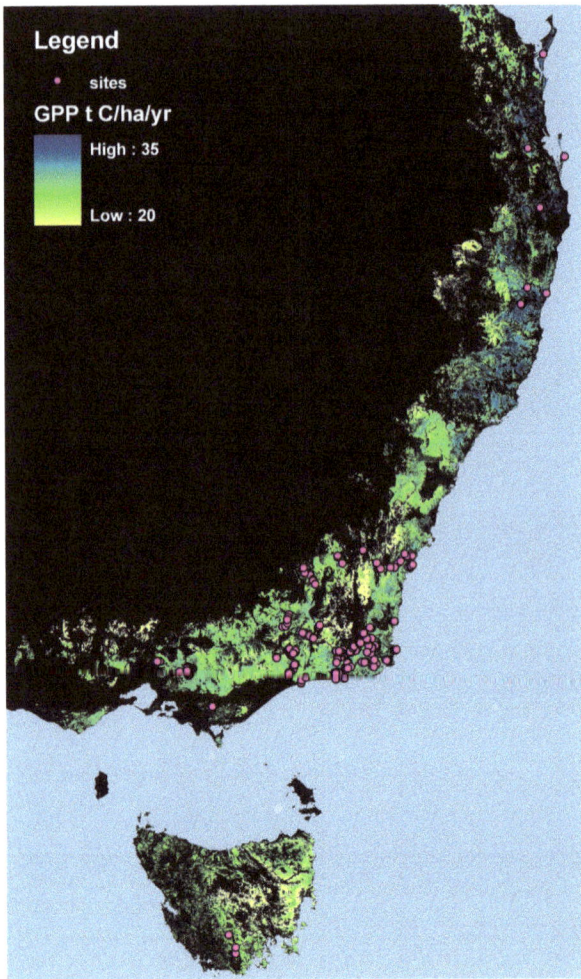

FIGURE 4: ESTIMATED GPP FOR THE STUDY REGION AND THE LOCATION OF FIELD SITES

The distribution of GPP by area is shown in the histogram, with a range of 12 to 33 t C ha^{-1} yr^{-1}.

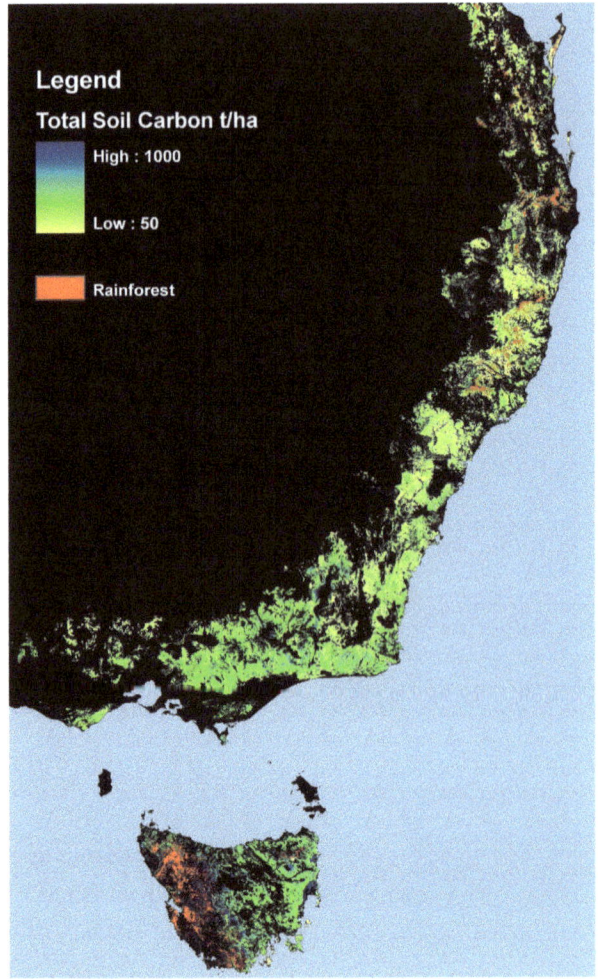

FIGURE 5: SPATIAL DISTRIBUTION OF TOTAL SOIL CARBON

The distribution of soil carbon by area is shown in the histogram, with a range of <50 to 2000 t C ha^{-1}.

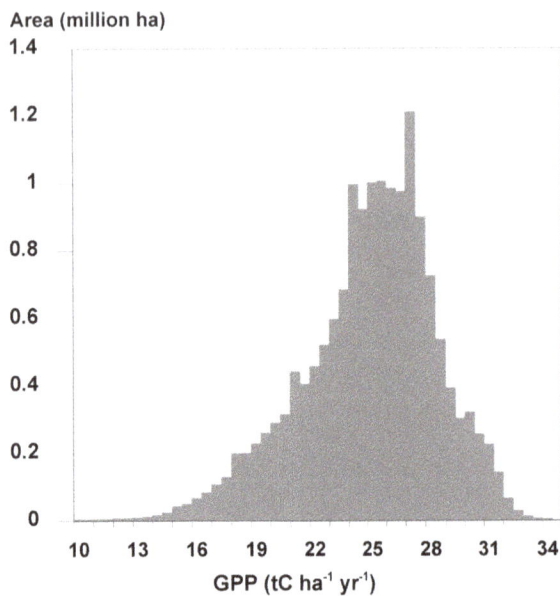

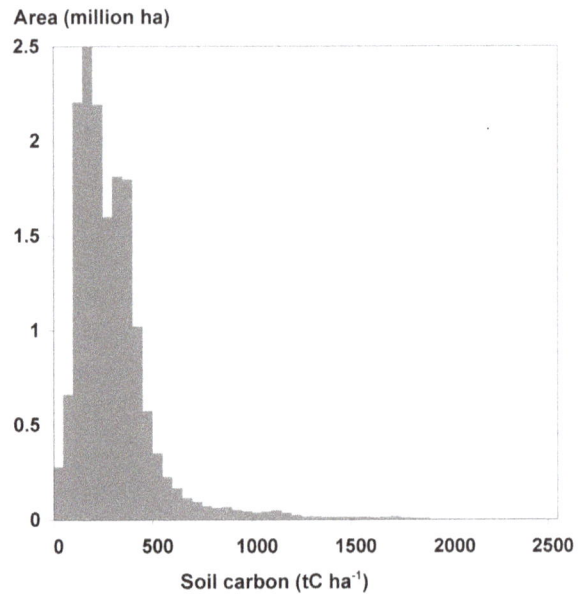

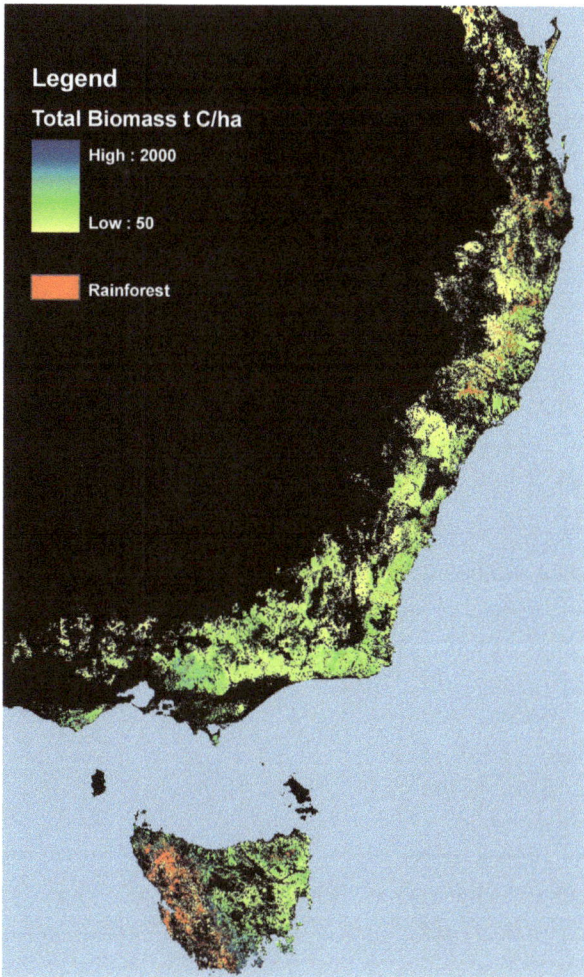

FIGURE 6: SPATIAL DISTRIBUTION OF THE TOTAL BIOMASS CARBON PREDICTED FROM THE MODEL

The distribution of total biomass carbon by area is shown in the histogram with a range of <50 to 2500 t C ha^{-1}.

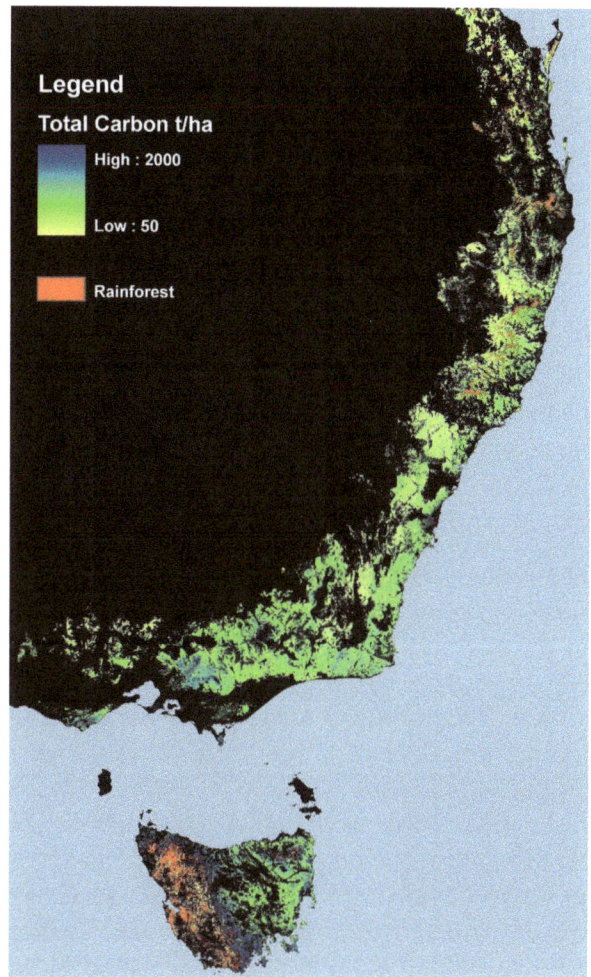

FIGURE 7: SPATIAL DISTRIBUTION OF TOTAL CARBON PREDICTED FROM THE MODEL (THAT IS, THE CARBON CARRYING CAPACITY)

The distribution of total carbon by area is shown in the histogram, with a range of <50 to 2500 t C ha^{-1}.

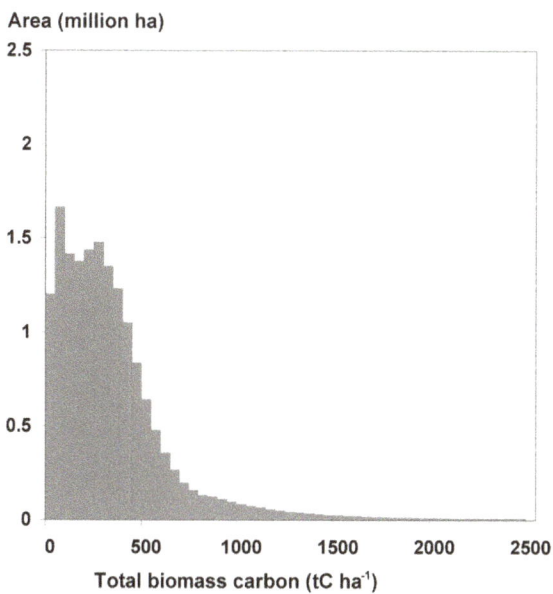

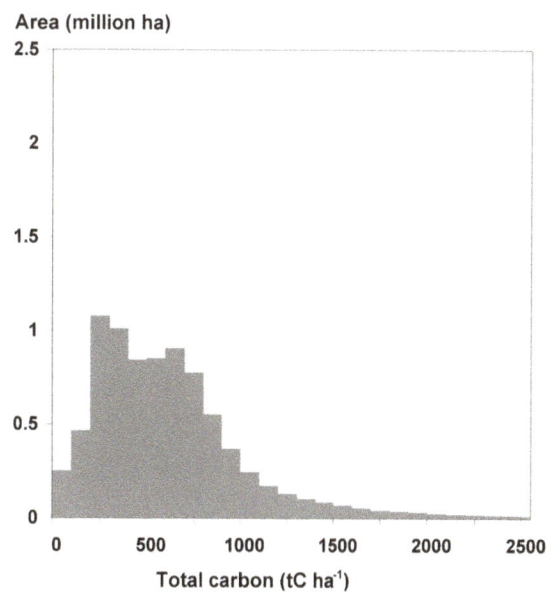

The proportion of carbon uptake used for biomass production is represented by the ratio of NPP:GPP. Relationships between GPP, NPP and biomass have been assumed to have constant coefficients in many modelling studies in the literature (for example, Waring et al. 1998). There has, however, been controversy about this issue (Keeling and Phillips 2007). We reviewed a global data set of 28 forest sites where NPP and GPP were measured and found that the ratio varied from 0.29 to 0.61. We statistically related NPP:GPP ratios with the corresponding environmental conditions for each site. This relationship improved the prediction of the proportion of carbon uptake used for biomass production compared with using a constant fraction of 0.47, which is used commonly in the literature. NPP was then estimated spatially by multiplying GPP for each grid cell in the GIS database by the NPP:GPP ratio predicted for that cell[11].

The living biomass carbon stock represents the balance between carbon accumulation from NPP and loss by mortality to the dead biomass carbon stock. The relationship between NPP and biomass carbon stock was investigated empirically using data from 240 sites in south-eastern Australia. These sites were in undisturbed mature forests and the data were collated from a range of sources and ecological studies. These field data were converted to spatial estimates of living biomass using appropriate allometric equations. Dead biomass includes the litter layer, coarse woody debris and standing dead trees. These components were measured only at some sites and, where there were no data, averages for forest types were used from a synthesis of information in the literature.

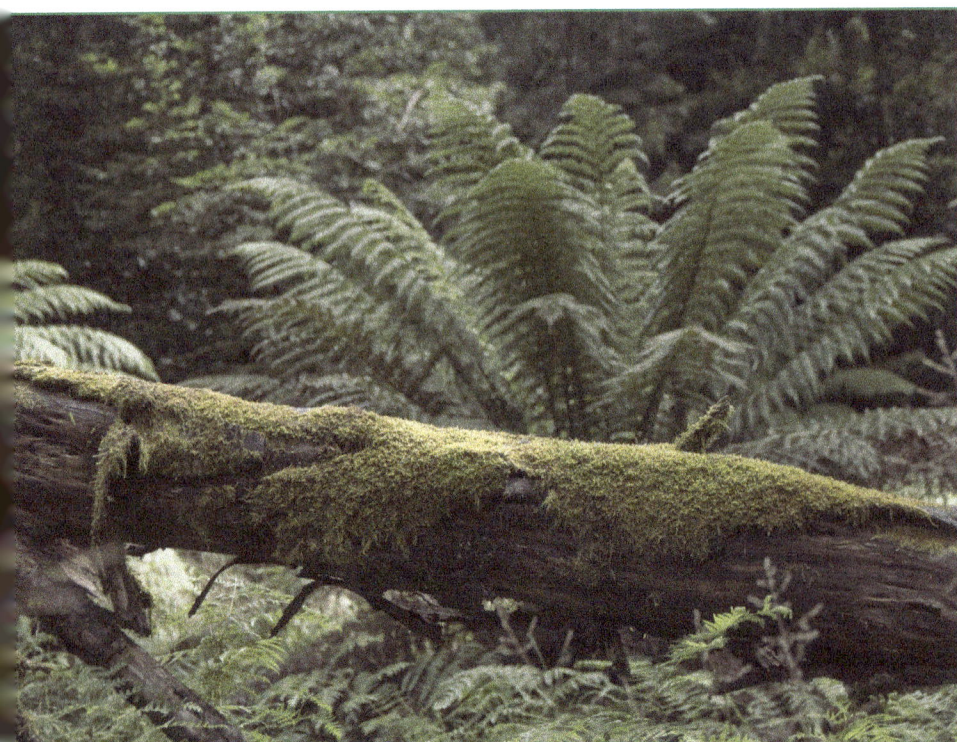

Coarse woody debris, central highlands, Victoria. Photo: Peter Halasz.

11   As noted in the introduction, full details of methods will be made available in a scientific paper currently in preparation.

The relationship between NPP and biomass stock is confounded by other factors that influence allocation and turnover rates. We investigated the use of environmental variables in conjunction with remotely sensed estimates of NPP as correlates to predict biomass. We used available spatial data for a selection of climatic, substrate and topographic environmental variables. A water availability index was also calculated and used as an ecologically meaningful expression of the interaction between precipitation and radiation. The effect of the environmental variables was described by a multiple regression model that accounted for 47 per cent of the variance in predicting total biomass in south-eastern Australian forests.

Soil carbon estimates were calculated from spatial data layers of soil depth, bulk density and soil carbon concentration as mapped by the Australian Soil Resource Information System (CSIRO 2007), and compared with site data where they existed. These values are for soil organic carbon only and would be higher if estimates of soil charcoal were available.

The analyses resulted in spatial predictions of living and dead biomass carbon and soil carbon, given prevailing environmental conditions, and assuming that the forests were ecologically mature and had not been disturbed by human activities. If the input field-site data have sampled landscape variability adequately, the effect of differences in climate, substrate, topography, wildfires and other natural disturbances should be reflected in these estimates. The statistical models enable the mean and standard deviation of carbon values to be calculated, where the latter can be interpreted in part to reflect the natural variability of conditions that affect forest growth in the region.

In this way, we were able to estimate and generate maps of the study region's natural carbon carrying capacity, thereby producing for the first time a baseline green carbon account for these natural forests.

RESULTS

The spatial distributions of the main components of the green carbon budget for the eucalypt forests of south-eastern Australia are shown for GPP (Figure 4), soil carbon (Figure 5), total biomass carbon (Figure 6) and total carbon (Figure 7) and are summarised in Table 1. Areas of rainforest are marked on these maps, but the carbon stock has not been predicted for them because there were insufficient site data from rainforests available for this study to predict biomass accurately. Predictions of carbon stocks have been made only within the numerical range of the input site data.

TABLE 1: SUMMARY OF THE CARBON STOCK OF EACH COMPONENT OF THE CARBON CARRYING CAPACITY OF THE EUCALYPT FORESTS OF SOUTH-EASTERN AUSTRALIA

Carbon component	Soil	Living biomass	Total biomass	Total carbon
Total carbon stock for the region (Mt C)	4060	4191	5220	9280
Carbon stock ha^{-1} (t C ha^{-1})	280	289	360	640
	(161)	(226)	(277)	(383)

Carbon stock per hectare is represented as a mean and standard deviation (in parentheses), which represents the variation in modelled estimates across the region. The study region covers an area of 14.5 million ha, representing 2 279 358 pixels at 250 m resolution.

Accumulation of carbon in biomass is related positively to NPP. Wide variance occurs, with many sites having a lower biomass for a given NPP than this maximum. This high spatial variability reflects the influence of environmental variables and natural disturbance regimes on the residence time of carbon in biomass components. The high spatial variability in carbon stocks across the region is represented as high standard deviations in Table 1, with particularly high values of carbon stocks covering only relatively small geographic areas.

The highest biomass carbon stocks (more than 1500 t C ha^{-1}) are in the mountain ash (*Eucalyptus regnans*) forest in the Central Highlands of Victoria (based on the forest types where data were available). This is cool temperate evergreen forest with a tall eucalypt overstorey and dense *Acacia* spp. and temperate-rainforest tree understorey. Environmental conditions are ideal for plant growth and accumulation of biomass, with high rainfall, moderate temperatures, moderately fertile and deep soils and in a sheltered valley. Highest biomass occurs in stands with two or three age cohorts of overstorey trees and rejuvenated understorey trees, which have resulted from partial stand-replacing wildfires (see Lindenmayer et al. 1999; Mackey et al. 2002).

Forest types where biomass is relatively low for a high NPP occur in the subtropics of northern coastal New South Wales and southern Queensland, where tree longevity is relatively lower and decomposition rates are higher than in temperate forests, resulting in lower accumulation of living and dead biomass. Sites with limiting environmental conditions—such as low water availability, infertile or shallow soils—also have lower biomass for a given NPP. Additionally, some forest stands might not be at maximum age and hence biomass, because the site history was uncertain.

COMPARISON WITH EXISTING CARBON ACCOUNTS

Soil matrix with fine roots: Red Dermosol, Brindabella Ranges. Photo: Heather Keith.

One way to understand the significance of our estimates of the carbon carrying capacity of the natural forests of south-eastern Australia is to compare them with values estimated from other sources. Two widely used sources of forest carbon data are the default values published by the Intergovernmental Panel on Climate Change (IPCC) and estimates derived from the Australian Government's National Carbon Accounting System (NCAS).

The IPCC recommends default values for estimating green carbon stocks in the absence of local data (Watson et al. 2001). Mean carbon stock and flux values are provided for the world's major biomes[12], as detailed in Table 2. Our analyses (Table 1) showed that the stock of carbon for intact natural forests in our study area is about 640 t C ha^{-1} and the average NPP of natural forests is 12 t C ha^{-1} yr^{-1} (with a standard deviation of 1.8). In terms of global biomes, Australian forests are classified as temperate forests. The IPCC default values for temperate forests are a carbon stock of 217 t C ha^{-1} and an NPP of 7 t C ha^{-1} yr^{-1}.

TABLE 2: ESTIMATED AVERAGE UPTAKE AND CARBON STOCKS IN THE WORLD'S MAIN FOREST BIOMES

Forest biome	NPP (t C ha^{-1} yr^{-1})	Carbon stock (t C ha^{-1})		
		Soil	Biomass	Total
Boreal forests	2.1	296	53	349
Temperate forests	7.0	122	96	217
Tropical forests	10.0	122	157	279

Source: Watson, R. T., Noble, I. R., Bolin, B., Ravindranath, N. H., Verardo, D. J. and Dokken, D. J. (eds) 2001, *Land Use, Land-Use Change, and Forestry*, Intergovernmental Panel on Climate Change (IPCC), Cambridge University Press, Third Assessment Report, Table 3.2.

Comparing the values in Tables 1 and 2, it can be seen that the IPCC default values represent only one-third of the natural carbon carrying capacity of the eucalypt forests of south-eastern Australia, and only 27 per cent of the biomass carbon stock. Using our figures, the total stock of carbon that can be stored in the 14.5 million ha of eucalypt forest in our study region is 9.3 Gt, if it is undisturbed by intensive human land-use activity and allowed to reach its natural carbon carrying capacity; applying the IPCC default values would give only 3.1 Gt. Note that while our model estimates the average total carbon stock of natural eucalypt forests at 640 t C ha^{-1}, real site values range up to 2500 t C ha^{-1}. This range reflects the natural variability found across landscapes in the environmental conditions and disturbance regimes that affect forest growth.

---

12   Biomes are large areas that have a similar climate and vegetation structure—that is, the vegetation has a similar height and density, even though the floristic composition might differ.

How can we explain the difference in total carbon between our estimates and the IPCC default values? The answer lies in the fact that current approaches to carbon accounting have been designed to estimate carbon stocks and flows in industrialized forests, including plantations. That is, they are designed to measure what we call *brown carbon*, not *green carbon*. As we discussed earlier, current approaches generally use field data from forestry mensuration plots. These plots are designed to provide estimates of growth rates in regenerating trees of commercial importance, which store much less carbon than unlogged natural forests. This is the main reason why carbon accounting methods that are calibrated using field data from industrialized forests significantly underestimate a landscape's carbon carrying capacity. There is also the problem of definition of forest and how different average values are compared. The definition of forest used in the Australian classification is trees taller than 10 m and canopy cover greater than 30 per cent, whereas the definition of forest used for the IPCC default values is trees taller than 2 m and canopy cover greater than 10 per cent (UNFCCC 2002). Additionally, the forests of south-eastern Australia have high GPP relative to typical default values.

Green carbon accounting tools for natural forests need to be calibrated using ecological field data obtained from sites that have not been disturbed by intensive human land-use activity, especially commercial logging. We made a special effort to find such ecological field data for our study region so that our estimates of carbon stocks were calibrated appropriately to represent the landscape's carbon carrying capacity.

Further insight into the requirements of green carbon accounting can be gained by comparing our estimates with those generated from the NCAS (Australian Greenhouse Office 2007a). The NCAS was designed to model biomass growth in plantations and afforestation/reforestation projects using native plantings. The empirically based calculations within the NCAS were calibrated using data appropriate for that purpose. Consequently, the NCAS was not designed to estimate the carbon carrying capacity of undisturbed natural forests.

To illustrate the need to calibrate carbon models using data that are appropriate for the purpose of a study, we used the NCAS to calculate carbon stocks at the locations for which we had obtained field data. Figure 8 shows the results of this analysis and compares the biomass estimates from the NCAS with our modelled predictions and with the real biomass calculated at each of the field sites used in our study (see Figure 4).

The NCAS generally underestimates biomass in natural forests that are largely undisturbed by human land-use activity—that is, the NCAS underestimates the carbon carrying capacity of natural forests. This is not surprising because it was not developed with this purpose in mind. The NCAS is a well-designed carbon accounting tool that represents the main ecological processes shown in

Figure 3. It is theoretically and technically possible to modify this program by calibrating it with data and empirical relationships—such as those we have used to develop our model—appropriate for the purpose of estimating the natural carbon carrying capacity of forests.

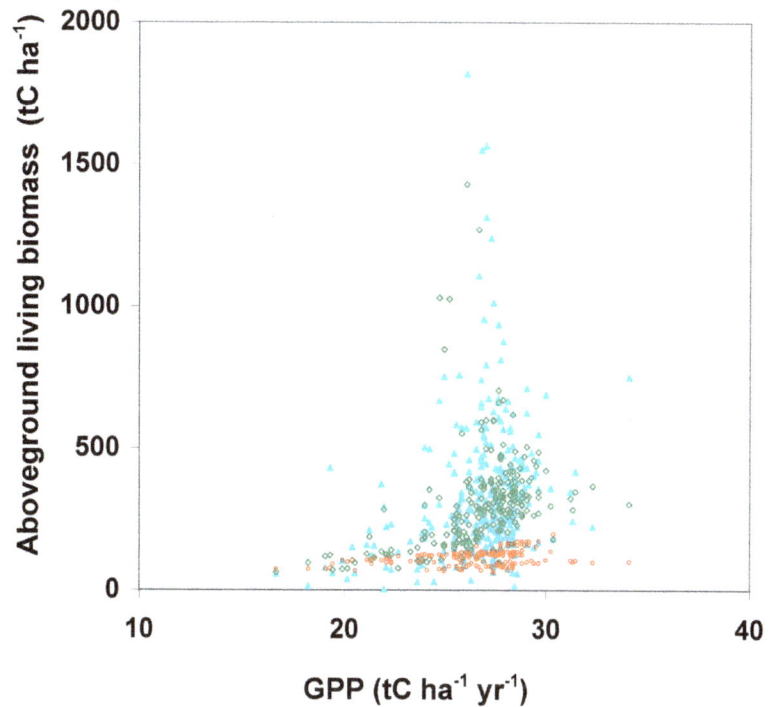

FIGURE 8: COMPARISON OF GPP AND BIOMASS

GPP was calculated by the methods used in this report and biomass estimates were derived from: i) the NCAS (orange open circles); ii) field sites (blue triangles); and iii) our modelled relationships between NPP and environmental variables (green open diamonds).

IMPLICATIONS FOR CARBON POLICY

THE IMPORTANCE OF CARBON CARRYING CAPACITY

We noted in the introduction that the Intergovernmental Panel on Climate Change (IPCC) has identified the need for forest-based mitigation analyses that account for natural variability in forest conditions, use primary forest structure and composition data and provide reliable baseline carbon accounts (Nabuurs et al. 2007). The approach we document in this study provides the means to generate such reliable baseline green carbon accounts for natural forests.

Once estimates of the carbon carrying capacity for a landscape have been derived, it is possible to calculate a forest's future carbon sequestration potential. This is the difference between a landscape's current carbon stock (under current land management) and the carbon carrying capacity (the maximum carbon stock when undisturbed by humans).

The current carbon stocks reflect the impact of human land-use activities in removing woody biomass from the forest, in some cases degrading soil carbon, and reducing residence time of organic carbon pools in the ecosystem. Some human activities also lead to an increase in fire, which again reduces current stocks, especially if there is post-fire salvage logging (Mackey et al. 2002).

The carbon sequestration potential is the amount of green carbon that potentially can be sequestered and stored in a landscape, if no further carbon-degrading land-use activity occurs and prevailing natural disturbance regimes persist. If a natural forest has not been subjected to intensive human land-use activity, the current carbon stock should be equal to the estimated carbon carrying capacity. When the carbon carrying capacity is known, the limiting factor in calculating the carbon sequestration potential of a landscape is the availability of data needed to calculate current carbon stocks, especially data about: 1) land-use history, and 2) the carbon stocks in dead and living woody biomass and soil. All of these data are needed on a landscape-wide basis.

The correct baseline to use when undertaking green carbon accounting is the carbon carrying capacity, against which the significance of changes in carbon stocks can be gauged. The calculation of most practical significance is the carbon sequestration potential. The approach developed by Roxburgh et al. (2006) includes a simulation model that, once calibrated properly, can estimate the carbon sequestration potential of natural forests. Such analyses are part of our continuing research activities.

Given the extensive impact of human land-use activities, particularly land clearing and all forms of commercial logging, carbon carrying capacity has to be estimated carefully in many landscapes from the best available data. If the carbon carrying capacity is not considered explicitly, the current carbon stock will be taken as representing the baseline against which future changes are gauged. Assuming there is a history of intensive land use, the result will be an underestimate of the green carbon account. The landscape's potential for carbon storage will have been undervalued.

DEFORESTATION AND FOREST DEGRADATION

After the 2007 Bali Climate Change Conference, the international community formally recognized the need to reduce emissions from deforestation and forest degradation as part of a comprehensive approach to solving the climate change problem. Deforestation is the result of a complex process reflecting the interaction of many factors such as national development priorities, local community needs and aspirations, the concerns of civil society organisations and commercial interests. Land and its resources are factors in production, and usually end up being allocated to the highest market-based economic value, unless governments intervene to protect non-market values through special conservation policies and legislation.

Clearing natural forests for bio-fuel plantations currently gives the highest economic return in many situations. Unfortunately, international rules defining forests and government carbon trading do not prevent natural forests in developing countries from being cleared for bio-fuel plantations. For example, in Indonesia, natural forests are being cleared for monoculture plantations of oil palms (Fargione et al. 2008). The international rules also do not prevent natural forests in developed countries being cleared for monoculture plantations (see Milne 2007).

Clearing natural forests to establish plantations does not reflect a scientific understanding of the difference between natural and industrialized forests. In terms of greenhouse gas emissions, the international rules that govern carbon trading and national-level policies do not distinguish between what we call in this report grey, brown and green carbon. Ignoring the difference between these forms of carbon can create ecologically perverse incentives for changing the land use and land cover.

It has now been shown that converting natural ecosystems to produce food-based bio-fuels creates a 'bio-fuel carbon debt' by releasing 17 to 420 times more carbon dioxide than the annual greenhouse gas reductions these bio-fuels provide by displacing fossil fuels (Fargione et al. 2008). The larger the natural carbon carrying capacity of a forest ecosystem (and the more intact the forest's carbon stocks), the greater will be the carbon debt from clearing to grow plantations. For eucalypt forests, recovery of the carbon debt from clearing intact natural forest through afforestation or reforestation takes more than 100 years (Roxburgh et al. 2006).

Forests are defined under the United Nations Framework Convention on Climate Change (UNFCCC) as woody vegetation of at least 2 m in height and 10 per cent canopy cover. It is therefore a simple structural definition based on the height and density of woody plants in an area (UNFCCC 2002)[13]. One reason for the perverse outcomes we are now witnessing in forests is the limitation of this definition and associated rules that do not reflect: 1) an understanding of green carbon accounting as presented here; and 2) an ecological and evolutionary scientific understanding about how a natural forest differs from an industrialized forest. To appreciate this difference, we need to consider the web of ecological and evolutionary processes that sustain the system within which the green carbon is stored.

In addition to the dominant tree canopy layer, natural forests contain a vast array of other plant species that support, through the biomass they produce from photosynthesis, an extraordinary diversity of animal species (mammals, birds, reptiles, invertebrates), fungi and a multitude of microbial organisms. A natural forest contains genetic information that is being copied continually (through reproduction), corrected (through the failure to survive of organisms with faulty copies), replaced (by the survivors) and revised (through proliferation of organisms possessing favourable modifications to the genome). Most importantly, this revision of the genome allows populations to adapt to environmental changes, including the climate change that we are currently experiencing.

Maintenance of the genetic diversity of natural forests, and therefore the capacity of the organisms contained therein to continue to adapt to environmental change, requires a self-perpetuating system. When land is deforested, this store of genetic information is reduced and the capacity of the remaining population of the species to adapt to environmental change is compromised. Clearing of natural forest reduces the population viability of the biota in the remaining unmodified forest (Lindenmayer and Fischer 2006). The living information in the genetic material of the forest biota regulates the bio-geochemical and ecosystem processes (Gorshkov et al. 2000). As natural forest is self-sustaining, it is able to persist without the need for management inputs from humans. Consequently, carbon accounting in natural forests need consider only the carbon gains and losses associated with biological processes; photosynthesis, respiration and oxidative combustion by wildfire and the production of charcoal.

In contrast with natural forests, industrialized forests comprise a very small number of species. Plantations are not self-sustaining systems; they consist of copies of genetic information and require a succession of energy inputs (mostly sourced from fossil fuels) during their lifetime, from seedling propagation to harvest. These include: site preparation (removal of existing vegetation), seed collection, growth trials to test the potential survival of species, seedling nursery inputs to grow seedlings for planting, planting of seedling

---

13   In addition to tree crown cover (>10-30%) and height (2-5 m) at maturity, the IPCC definition of forest includes consideration of the minimum area (0.05-1.0 ha) and width of land.

trees, application of herbicides to suppress competition from weed species, measures to prevent animal species (vertebrates and invertebrates) from browsing on the seedlings, fertilizer application (most soils in Australia are nutrient impoverished) and continuing maintenance to suppress plant and animal pest species and fire.

As plantations are not self-sustaining systems, when the trees are harvested or die, energy inputs (again, sourced mostly from fossil fuels) are required to establish a new crop of trees. All of these fossil-fuel inputs, including those required for the manufacture of consumables such as fertilizer and pesticides, need to be taken into account, along with the biological processes, when assessing the carbon sequestration potential of tree plantations (and other agricultural crops). As plantations are eventually harvested, the fossil-fuel inputs, such as those required for road-making and upgrading, transport of the saw-logs for processing, the energy needs (and carbon dioxide emissions) for processing of timber or woodchips, and other industrial processes, should also be deducted from the gross pre-harvest carbon stock.

Despite the progress we are now seeing in the development of international policy responses to the problem of deforestation, there remains a lack of clarity about the kinds of human activities that contribute to forest degradation. From a climate change perspective, forest degradation needs to be defined to include the impact of all human land-use activity that reduces the current carbon stock in a natural forest compared with its natural carbon carrying capacity. The impact of commercial logging on natural forests must therefore also be considered when accounting for forest degradation. As discussed earlier, commercially logged forests have substantially lower carbon stocks and reduced biodiversity than intact natural forests, and studies have shown carbon stocks to be 40 to 60 per cent lower depending on the intensity of logging (Brown et al. 1997; Dean et al. 2003; Roxburgh et al. 2006). In Brazilian Amazon, the area of natural forest that is logged commercially resulting in degraded carbon stocks is equivalent to that subject to deforestation and represents approximately 0.1 Gt of green carbon emissions to the atmosphere (Asner et al. 2005).

While clearing for agriculture (either intensive or subsistence) can be a major cause of deforestation and forest degradation (especially in tropical forests), commercial logging can also be the initial causal factor. Depending on the prevailing regulatory framework, a succession of planned and unplanned, legal and illegal land-use activities can be introduced into a landscape facilitated by the logging infrastructure—in particular, the road network. The end point of this process can be broad-scale degradation and deforestation, with associated increased carbon dioxide emissions.

GREEN CARBON AND MITIGATION

Given the scale and urgency of the climate change problem, we need to take a fresh look at the contribution natural forests can make to mitigating rising levels of atmospheric carbon dioxide. We can illustrate the implications of taking a fresh approach by considering again the carbon carrying capacity we have calculated for the eucalypt forests in south-eastern Australia (Figure 7). Our comments here, however, can be of a preliminary nature only as we have not yet calculated the carbon sequestration potential of these forests—a task that remains part of our continuing research.

About 30 per cent of Australia's forests have been cleared and the land converted to agricultural or other land uses. Of the 14.5 million ha of eucalypt forest shown in Figure 7 (which is about half of Australia's remaining eucalypt forests), about 4.9 million ha are in some kind of protected area, while 9.6 million ha are on either public or private land. Of the unprotected natural forest, about 8.1 million ha (about 56 per cent) have been logged commercially.

Protecting natural forests can be part of a comprehensive mitigation strategy in two ways:

1. keeping the carbon in the forest ecosystem—that is, in the biomass and bound to soil particles

2. allowing the forests that have been logged previously to re-grow and reach their carbon sequestration potential.

The carbon carrying capacity of the 14.5 million ha of eucalypt forest in our study area is about 9 Gt C (equivalent to 33 Gt CO_2). About 44 per cent of the area has not been logged and can be considered at carbon carrying capacity, which represents about 4 Gt C (equivalent to 14.5 Gt CO_2). About 56 per cent of the area has been logged, which means these forests are substantially below their carbon carrying capacity of 5 Gt C. If it is assumed that logged forest is, on average, 40 per cent below carbon carrying capacity (Roxburgh et al. 2006), the current carbon stock is 3 Gt C (equivalent to 11 Gt CO_2). The total current carbon stock of the 14.5 million ha is 7 Gt C (equivalent to 25.5 Gt CO_2). If logging in native eucalypt forests was halted, the carbon stored in the intact forests would be

E. regnans in Mt. Baw Baw, Victoria. Photo: Chris Taylor.

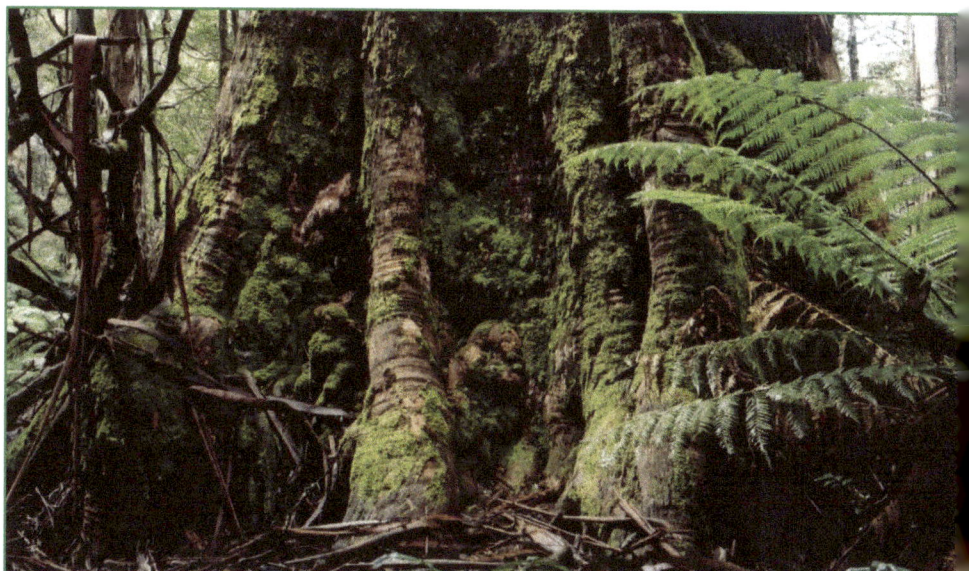

protected and the degraded forests would be able to regrow their carbon stocks to their natural carbon carrying capacity. Based on the assumptions above, the carbon sequestration potential of the logged forest area is 2 Gt C (equivalent to 7.5 Gt CO_2).

Costa and Wilson (2000) have derived an equivalence factor to relate the stock of carbon in the biosphere to the effect of the emitted carbon dioxide in the atmosphere, stated as "the effect of keeping 1 t CO_2 out of the atmosphere for 1 year". This is based on the inference that "removing 1 t CO_2 from the atmosphere and storing it for 55 years counteracts the radiative forcing effect, integrated over a 100 year time horizon, of a 1 t CO_2 emission pulse". Applying this equivalence factor, every 1 t CO_2 sequestered as a biosphere stock for 55 years is equal, in a radiative forcing context, to 0.0182 t CO_2 yr^{-1} (for 100 years) of avoided emissions, and every 1 Gt CO_2 stored is equivalent to 18.2 Mt CO_2 yr^{-1} (for 100 years) of avoided emissions. The effect of retaining the current carbon stock of 25.5 Gt CO_2 in our study area is therefore equivalent to avoided emissions of 460 Mt CO_2 yr^{-1} for the next 100 years. Allowing logged forests to realize their sequestration potential to store 7.5 Gt CO_2 is equivalent to avoiding emissions of 136 Mt CO_2 yr^{-1} for the next 100 years. This amount of emissions is equal to 24 per cent of the 2005 Australian net greenhouse gas emissions across all sectors (559 Mt CO_2 yr^{-1}) (Australian Greenhouse Office 2007b). This approach is assuming a 100 year lifetime for most of the carbon dioxide in the atmosphere. However, Archer (2005) considers a better approximation of the lifetime of fossil fuel carbon dioxide might be "300 years plus 25% that lasts forever".

Another way of appreciating the relative importance of the carbon stock in forests is to compare it with the stock in the atmosphere. If the entire carbon stock was released from the forests in our study area into the atmosphere, it would raise the global concentration of carbon dioxide by 3.3 ppmv[14]. This is a globally significant amount of carbon dioxide; since 1750 AD, the concentration of carbon dioxide in the atmosphere has increased by some 97 ppmv.

It is possible to achieve protection of the carbon stocks in natural forests by switching to timber sourced from existing plantations and, if necessary, from new plantations on previously cleared land. In this way, the commercial demand for wood fibre can be met and the contribution of natural forests to greenhouse gas mitigation can be maximized. Currently, about 68 per cent of wood fibre is sourced from the plantation estate, but current plantation stocks are sufficient to meet nearly all the national demand for wood and paper products (Ajani 2007).

---

14   1 ppmv CO_2 in the atmosphere is equivalent to 2.13 Gt C (Carbon Dioxide Information Analysis Center).

CONCLUDING COMMENTS

Canopy leaves: *E. delegatensis*, Bago State Forest, southern NSW. Photo: Heather Keith.

In considering the role of natural forests in the climate change problem, we must avoid the temptation to take a reductionist approach in which all we see is a measure of carbon with a fungible, market value. Much of what distinguishes natural forests from industrialized forests cannot be measured let alone assigned a market value. We are just beginning to understand the powerful ways in which micro-evolutionary processes enable local adaptations in very dynamic ways and over what were previously considered to be ecological time scales (Bradshaw and Holzapfel 2006). Molecular analyses are also revealing the extraordinary complexity, persistence and geographic patterning of coevolutionary relationships between populations and across communities (Thompson 2005). Indeed, it is these elusive biological, ecological and evolutionary attributes that underpin the qualities that make green carbon in natural forests a more reliable and resilient stock compared with the brown carbon of industrialized forests. Green carbon is not analogous to the grey carbon of coal; it emerges from and is part of complex, adaptive ecosystems.

Carbon accounting models must be calibrated specifically with appropriate ecological field data before they can generate reliable estimates for natural forests. Default Intergovernmental Panel on Climate Change (IPCC) values and accounting tools developed for industrialized forests will not generate reliable estimates for natural forests. Green carbon accounting for natural forests is needed, based on reliable estimates of: 1) the carbon carrying capacity; 2) current carbon stocks; and 3) the carbon sequestration potential. With these data, it is possible to evaluate the carbon uptake from, or emission to, the atmosphere from changing land-use activities and land cover. Our approach to green carbon accounting enables these essential calculations to be undertaken. It addresses the IPCC's call for the need for forest-based mitigation analyses that account for natural variability, use primary data and provide reliable baseline carbon accounts.

Forest degradation should be defined from a climate change perspective to include any human land-use activity that reduces the carbon stocks of a forested landscape relative to its carbon carrying capacity. The climate change imperative demands that we take a fresh look at our forest estate. The carbon impacts of all land uses, including commercial logging, must be brought explicitly into our calculations in terms of their direct and indirect effects on forest degradation.

The remaining intact natural forests constitute a significant standing stock of carbon that should be protected from carbon-emitting land-use activities. There is substantial potential for carbon sequestration in forest areas that have been logged commercially, if allowed to regrow undisturbed by further intensive human land-use activities.

As the world community begins the difficult and complex task of negotiating the terms for the post-2012 commitment period under the United Nations Framework Convention on Climate Change (UNFCCC), various mechanisms are being proposed to provide the incentives and investments necessary for forest protection, particularly in developing countries. The international regulatory framework being developed to help reduce emissions from deforestation and degradation needs to be based on a scientific understanding of natural forests and the ecological differences between natural forests and industrialized forests, especially monoculture plantations. Protecting existing natural forests from deforestation is important because it prevents the increase in atmospheric carbon dioxide levels that will necessarily result. The imperative to protect what is left of the world's natural forests (in addition to their intrinsic and other non-market values) comes from recognising their role in the global carbon cycle and the need to keep intact an essential component of Earth's life-support systems. The green carbon stored in natural forests is a significant component of the global carbon cycle, and about 18 per cent of annual greenhouse gas emissions come from deforestation.

More reliable estimates of baseline green carbon will enable the contribution of natural forests to the global carbon cycle to be valued properly. Our analyses show that in Australia and probably globally, the carbon carrying capacity of natural forests is underestimated and therefore misrepresented in economic valuations and in policy options. Scientifically, it is important to reduce emissions from deforestation and forest degradation in all forest biomes— boreal, tropical and temperate—and in economically developed as well as developing countries. Green carbon accounting and forest protection of all natural forests in all nations must become part of a comprehensive approach to the climate change problem.

E. regnans in Styx valley, Tasmania (1300 t C ha⁻¹ of biomass carbon). Photo: Geoff Law.

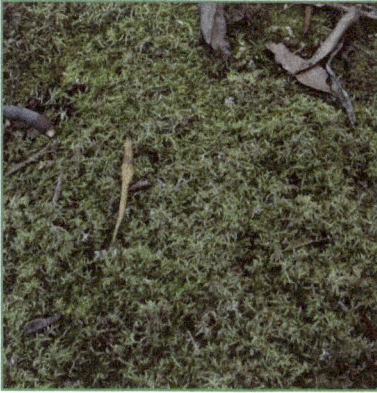

ACKNOWLEDGMENTS

Moss on fallen log: *E. obliqua*,
Mt Wellington, Tasmania. Photo: Rob
Blakers.

Natural forest of multi-aged *E. regnans*
with rainforest understorey in the
O'Shannassy catchment, central highlands,
Victoria. (1800 t C ha⁻¹ of biomass
carbon). Photo: Luke Chamberlain.

The provision of unpublished data from Andrew Claridge, Phil Gibbons, Kris Jacobsen and Charlie Mackowski is gratefully acknowledged. In collaboration with these researchers, a technical paper is in preparation, detailing methods and results, for submission to a scientific journal. Any errors in this report, however, and the comments on the results' policy implications, are solely the responsibility of Mackey, Keith, Berry and Lindenmayer.

We are grateful to The Wilderness Society Australia for a research grant that supported the analyses presented in this report. Thanks also to Clive Hilliker, Luciana Porfirio and Andrew Wong for their assistance in the technical production of this report, and Michael Roderick and Henry Nix for helpful comments on a draft. We thank the photographers, Sarah Rees, Luke Chamberlain, Rob Blakers, Chris Taylor, Ern Mainka, Geoff Law, Andrew Wong, Ian Smith, Michael Hodda and Peter Halasz, for the use of their images. These analyses also drew on data and models developed as part of an Australian Research Council Linkage grant, LP0455163.

REFERENCES

Fungi in the litter layer: *E. fastigata* forest, Shoalhaven catchment. Photo: Sandra Berry.

Corymbia maculata, south coast NSW. Photo: Andrew Wong.

Abe, H., Sam, N., Niangu, M., Vatnabar, P. and Kiyono, Y. 1999, 'Effect of logging on forest structure at the Mongi-Busiga forest research plots, Finschhafen, Papua New Guinea', Proceedings of the PNGFRI–JICA International Forestry Seminar, 4–7 October 1999, *PNGFRI Bulletin No. 18*, Papua New Guinea Forest Research Institute.

Ajani, J. 2007, *The Forest Wars*, Melbourne University Press, Carlton.

Archer, D. 2005, 'Fate of fossil fuel CO_2 in geologic time', *Journal of Geophysical Research*, 110, pp. 1–6.

Asner, G., Knapp, D. E., Broadbent, E. N., Oliveira, P. J. C., Keller, M. and Silva, J. N. 2005, 'Selective logging in the Brazilian Amazon', *Science*, 310, pp. 480–2.

Australian Greenhouse Office 2007a, *National Carbon Accounting System*, Australian Greenhouse Office, Department of the Environment, Water, Heritage and the Arts, <http://www.greenhouse.gov.au/ncas >

Australian Greenhouse Office, 2007b, *National Greenhouse Gas Inventory*, Australian Greenhouse Office, Department of the Environment, Water, Heritage and the Arts, <http:/www.greenhouse.gov.au/inventory>

Barrett, D. J. 2002, 'Steady state turnover time of carbon in the Australian terrestrial biosphere', *Global Biogeochemical Cycles*, 16 (4), p. 1108, doi:10.1029/2002GB001860.

Barrett, D. J., Byrne, G. and Senarath, U. 2005, 'The 250m 16-day vegetation index product (MOD13Q1) for Australia', *CSIRO Land and Water Technical Report*, 05/05, CSIRO Land and Water, Canberra.

Berger, A. and Loutre, M. F. 2002, 'CLIMATE: An exceptionally long interglacial ahead?', *Science*, 297, pp. 1287–8.

Berry, S., Mackey, B. and Brown, T. 2007, 'Potential applications of remotely sensed vegetation greenness to habitat analysis and the conservation of dispersive fauna', *Pacific Conservation Biology*, 13, pp. 120–7.

Brack, C. L., Richards, G. and Waterworth, R. 2006, 'Integrated and comprehensive estimation of greenhouse gas emissions from land systems', *Sustainability Science*, 1, pp. 91–106.

Bradshaw, W. E. and Holzapfel, C. M 2006, 'Evolutionary response to rapid climate change', *Science*, 312, pp. 147–8.

Brown, J. H., Gillooly, J. F., Allen, A. P., van Savage, M. and West, G. B. 2004, 'Toward a metabolic theory of ecology', *Ecology*, 85, pp. 1771–89.

Brown, S., Schroeder, P. and Birdsey, R. 1997, 'Aboveground biomass distribution of US eastern hardwood forests and the use of large trees as an indicator of forest development', *Forest Ecology and Management*, 96, pp. 37–47.

Costa, P. M. and Wilson, C. 2000, 'An equivalence factor between CO_2 avoided emissions and sequestration—description and applications in forestry', *Mitigation and Adaptation Strategies for Global Change*, 5, pp. 51–60.

Cowan, I. R. and Farquhar, G. D. 1977, 'Stomatal function in relation to leaf metabolism and environment', *Symposia of the Society for Experimental Biology*, 31, pp. 471–505.

CSIRO 2007, *Australian Soil Resource Information System*, <http://www.asris.csiro.au/index_ie.html>

Dean, C., Roxburgh, S. and Mackey, B. 2003, 'Growth modelling of *Eucalyptus regnans* for carbon accounting at the landscape scale', in A. Amaro, D. Reed and P. Soares (eds), *Modelling Forest Systems*, CABI, Oxon, pp. 27–39.

Des Marais, D. J. 2000, 'Evolution: When did photosynthesis emerge on Earth?', *Science*, 289, pp. 1703–5.

Fargione, J., Hill, J., Tilman, D., Polasky, S. and Hawthorne, P. 2008, 'Land clearing and the biofuel carbon debt', *Science*, 319, pp. 1235–8.

Farquhar, G. D. 1997, 'Climate change: Carbon dioxide and vegetation', *Science*, 21, p. 1411.

Farquhar, G. D. and Roderick, M. L. 2003, 'Pinatubo, diffuse light, and the carbon cycle', *Science*, 299, pp. 1997–8.

Golley, F. B. 1983, 'Decomposition', in F. B. Golley (ed.), *Ecosystems of the World 14A—Tropical Rainforest Ecosystems—Structure and Function*, Elsevier Scientific Publishing Company, Amsterdam, pp. 157–66.

Gorshkov, V. G., Gorshkov, V. V. and Makarieva, A. M. 2000, *Biotic Regulation of the Environment: Key issues of global change*, Springer Praxis Books, UK.

Gupta, R. K. and Rao, D. L. N. 1994, 'Potential of wastelands for sequestering carbon by reforestation', *Current Science*, 66, pp. 378–80.

Hansen, J., Sato, M., Kharecha, P., Russell, G., Lea, D. W. and Siddall, M. 2007, 'Climate change and trace gases', *Philosophical Transactions of the Royal Society* A, 365, pp. 1925–54.

Harmon, M. E., Ferrell, W. K. and Franklin, J. F. 1990, 'Effects on carbon storage of conversion of old-growth forests to young forests', *Science*, 247, pp. 699–702.

Hollinger, D. Y., Kelliher, F. M., Schulze, E.-D., Bauer, G., Arneth, A., Byers, J. N., Hunt, J. E., McSeveny, T. M., Kobak, K. I., Milukova, I., Sogatchev, A., Tatarinov, F., Varlargin, A., Ziegler, W. and Vygodskaya, N. N. 1998, 'Forest–atmosphere carbon dioxide exchange in eastern Siberia', *Agricultural and Forest Meteorology*, 90, pp. 291–306.

Hooper, D. U. and Vitousek, P. M. 1997, 'The effects of plant composition and diversity on ecosystem processes', *Science*, 277, pp. 1302–5.

Hooper, D. U., Chapin, F. S., Ewel, J. J., Hector, A., Inchausti, P., Lavorel, S., Lawton, J. H., Lodge, D. M., Loreau, M., Naeem, S., Schmid, B., Setala, H., Symstad, A. J., Vandermeer, J. and Wardle, D. A. 2005, 'Effects of biodiversity on ecosystem functioning: a consensus of current knowledge', *Ecological Monographs*, 75, pp. 3–35.

Houghton, R. 2007, Balancing the global carbon budget. *Annual Review of Earth and Planetary Sciences*, 35, pp. 313 - 347.

Intergovernmental Panel on Climate Change (IPCC) 2007, *The Fourth Assessment Report Climate Change 2007: Synthesis report*, Intergovernmental Panel on Climate Change, <http://www.ipcc.ch/>

Keeling, H. C. and Phillips, O. L. 2007, 'The global relationship between forest productivity and biomass', *Global Ecology and Biogeography*, 16, pp. 618–31.

Lindenmayer, D. B., Incoll R. D., Cunningham, R. B. and Donnelly, C. F. 1999, 'Attributes of logs on the floor of Australian mountain ash (*Eucalyptus regnans*) forests of different ages', *Forest Ecology and Management*, 123, pp. 195–203.

Lindenmayer D.B. and Fischer J. 2006, *Habitat fragmentation and landuse change: An ecological and conservation synthesis.* Island Press, Washington, USA.

McCann, K. 2007, 'Protecting biostructure', *Nature*, 446, p. 29.

Mackey, B., Berry, S. and Brown, T. 2008, 'Reconciling approaches to biogeographic regionalization: a systematic and generic framework examined with a case study of the Australian continent', *Journal of Biogeography*, 35, pp. 213–29.

Mackey, B. G., Lindenmayer, D. B., Gill, A. M., McCarthy, A. M. and Lindesay, J. A. 2002, *Wildlife, Fire and Future Climate: A forest ecosystem analysis*, CSIRO Publishing.

Millennium Ecosystem Assessment (MEA) 2005, *Ecosystems and Human Well-Being: Biodiversity synthesis*, World Resources Institute, Washington, DC.

Milne, C. 2007, Land clearing on the Tiwi Islands, Speech to the Australian Senate, <http://www.christinemilne.org.au/500_parliament_sub.php?deptItemID=163>

Nabuurs, J-G. et al. 2007, 'Forestry', in E. Metz, O. R. Robinson, P. R. Bosch, R. Dave and L. A. Meyer (eds), *Climate Change 2007: Mitigation. Contribution of Working Group III to the Fourth Assessment Report of the IPCC*, Cambridge University Press, Cambridge.

NVIS 2003, National Vegetation Information System, version 6.0. <http://www.environment.gov.au/erin/nvis>

Odum, E. P. and Barret, G. W. 2004, *Fundamentals of Ecology*, Thomson Brooks/Cole, Belmont, CA.

Odum, H. T. 1981, *Energy Basis for Man and Nature*, McGraw-Hill Book Company, New York.

Petit, J. R., Jouzel, J., Raynaud, D., Barkov, N. I., Barnola, J.-M., Basile, I., Bender, M., Chappellaz, J., Davis, M., Delaygue, G., Delmotte, M., Kotlyakov, V. M., Legrand, M., Lipenkov, V. Y., Lorius, C., Pépin, L., Ritz, C., Saltzman, E. and Stievenard, M. 1999, 'Climate and atmospheric history of the past 420,000 years from the Vostok ice core, Antarctica', *Nature*, 399, pp. 429–36.

Roderick, M. L., Farquhar, G. D., Berry, S. L. and Noble, I. R. 2001, 'On the direct effect of clouds and atmospheric particles on the productivity and structure of the vegetation', *Oecologia*, 129, pp. 21–31.

Roxburgh, S. H., Wood, S. W., Mackey, B. G., Woldendorp, G. and Gibbons, P. 2006, 'Assessing the carbon sequestration potential of managed forests: a case study from temperate Australia', *Journal of Applied Ecology*, 43, pp. 1149–59.

Saldarriaga, J. G., West, D. C., Thorpe, M. L. and Uhl, C. 1988, 'Long-term chronosequence of forest succession in the upper Rio Negro of Colombia and Venezuela', *Journal of Ecology*, 76, pp. 938–58.

Schulze, E.-D., Wirth, C. and Heimann, M. 2000, 'Managing forests after Kyoto', *Science*, 289, pp. 2058–9.

Thompson, J. N. 2005, *The Geographic Mosaic of Coevolution*, The University of Chicago Press, Chicago.

United Nations Framework Convention on Climate Change (UNFCCC),<http://unfccc.int/resource/docs/convkp/conveng.pdf>

United Nations Framework Convention on Climate Change (UNFCCC) 2002, *Biome-Specific Forest Definitions—Technical paper*, Report to the UNFCCC Secretariat, <http://www.fao.org/docrep/005/y417e/Y4171E52.htm>

Varmola, M. and Del Lungo, A. 2003, 'Planted forests database (PFDB): structure and contents', *Planted Forests and Trees Working Papers*, 25, Forest Resources Development Service, Forest Resources Division, Food and Agriculture Organisation, Rome.

Waring, R. H., Landsberg, J. J. and Williams, M. 1998, 'Net primary production of forests: a constant fraction of gross primary production?', *Tree Physiology*, 18, pp. 129–34.

Watson, R. T., Noble, I. R., Bolin, B., Ravindranath, N. H., Verardo, D. J. and Dokken, D. J. (eds) 2001, *Land Use, Land-Use Change, and Forestry*, Intergovernmental Panel on Climate Change (IPCC), Cambridge University Press.

Zhang, X, Zwiers, F. W., Hegerl, G. C., Lambert, F. H., Gillet, N. P., Solomon, S., Stott, P. A. and Nozwa, T. 2007, 'Detection of human influence on twentieth-century precipitation trends', *Nature*, 448, pp. 461–5.

Corymbia maculata, south coast NSW (580 t C ha^{-1} of biomass carbon). Photo: Sandra Berry.

E. regnans, East Gippsland (700 t C ha^{-1} of biomass carbon). Photo: Ern Mainka. (Overleaf)